TUMORS
and
CANCERS

HEAD—NECK—HEART—LUNG—GUT

T0144569

POCKET GUIDES TO
BIOMEDICAL SCIENCES

https://www.crcpress.com/Pocket-Guides-to-Biomedical-Sciences/book-series/
CRCPOCGUITOB

The *Pocket Guides to Biomedical Sciences* series is designed to provide a concise, state-of-the-art, and authoritative coverage on topics that are of interest to undergraduate and graduate students of biomedical majors, health professionals with limited time to conduct their own searches, and the general public who are seeking for reliable, trustworthy information in biomedical fields.

TUMORS
and
CANCERS
HEAD—NECK—HEART—LUNG—GUT

Dongyou Liu

CRC Press
Taylor & Francis Group
Boca Raton London New York

CRC Press is an imprint of the
Taylor & Francis Group, an **informa** business

CRC Press
Taylor & Francis Group
6000 Broken Sound Parkway NW, Suite 300
Boca Raton, FL 33487-2742

© 2018 by Taylor & Francis Group, LLC
CRC Press is an imprint of Taylor & Francis Group, an Informa business

No claim to original U.S. Government works

Printed on acid-free paper

International Standard Book Number-13: 978-1-138-08838-2 (Paperback)
978-1-138-30085-9 (Hardback)

Library of Congress Cataloging-in-Publication Data

Names: Liu, Dongyou, author.
Title: Tumors and cancers : head-neck-heart-lung-gut / Dongyou Liu.
Description: Boca Raton : Taylor & Francis, 2018. | Series: Pocket guides to biomedical sciences | "A CRC title, part of the Taylor & Francis imprint, a member of the Taylor & Francis Group, the academic division of T&F Informa plc." | Includes bibliographical references and index.
Identifiers: LCCN 2017011622 | ISBN 9781138088382 (paperback : alk. paper) : 9781138300859 (hardback)
Subjects: LCSH: Tumors. | Cancer--Popular works. | Head--Cancer. | Neck--Cancer. | Heart--Cancer. | Lung--Cancer. | Gastrointestinal system--Cancer.
Classification: LCC RC261 .L675 2018 | DDC 616.99/4--dc23
LC record available at https://lccn.loc.gov/2017011622

Visit the Taylor & Francis Web site at
http://www.taylorandfrancis.com

and the CRC Press Web site at
http://www.crcpress.com

Contents

SECTION III   Digestive System

Series Preface

Dramatic breakthroughs and nonstop discoveries have rendered biomedicine increasingly relevant to everyday life. Keeping pace with all these advances is a daunting task, even for active researchers. There is an obvious demand for succinct reviews and synthetic summaries of biomedical topics for graduate students, undergraduates, faculty, biomedical researchers, medical professionals, science policy makers, and the general public.

Recognizing this pressing need, CRC Press has established the *Pocket Guides to Biomedical Sciences* series, with the main goal to provide state-of-the-art, authoritative reviews of far-ranging subjects in short, readable formats intended for a broad audience. Volumes in the series will address and integrate the principles and concepts of the natural sciences and liberal arts, especially those relating to biomedicine and human well-being. Future volumes will come from biochemistry, bioethics, cell biology, genetics, immunology, microbiology, molecular biology, neuroscience, oncology, parasitology, pathology, and virology, as well as other related disciplines.

Forming part of the four volumes devoted to human tumors and cancers in the series, the current volume focuses on the head, neck, cardiovascular, respiratory, and digestive systems. Characterized by uncontrolled growth of abnormal cells that often extend beyond their usual boundaries and disrupt the normal functions of affected organs, tumors and cancers are insidious diseases with serious consequences. Relative to our ongoing research and development efforts, our understanding of tumors and cancers remains rudimentary, and the arsenal at our disposal against these increasingly prevalent diseases is severely limited. The goal of this volume is the same as the goal for the series—to simplify, summarize, and synthesize a complex topic so that readers can reach to the core of the matter without the necessity of carrying out their own time-consuming literature searches.

We welcome suggestions and recommendations from readers and members of the biomedical community for future topics in the series and experts as potential volume authors/editors.

Dongyou Liu
Sydney, Australia

Contributors

Jayantha Balawardena
Department of Oncology
Kotelawala Defence University
Colombo, Sri Lanka

Raúl Barrera-Rodríguez
Department of Biochemistry and
    Environmental Medicine
National Institute of Respiratory
    Diseases
Mexico City, Mexico

Megan M. Boniface
Esophageal & Gastric
    Multidisciplinary Clinic
Division of Surgical Oncology
Department of Surgery
University of Colorado Denver
Denver, Colorado

Kemal I. Deen
Department of Surgery
University of Kelaniyaand
Asiri Surgical Hospital
Colombo, Sri Lanka

Andreas V. Hadjinicolaou
MRC Human Immunology Unit
Weatherall Institute of Molecular
    Medicine
and
Radcliffe Department of Medicine
University of Oxford
Oxford, UK

Christopher Hadjittofi
General Surgery
Whipps Cross University Hospital
Barts Health NHS Trust
London, UK

Zainul A. Kapacee
Manchester Royal Infirmary
Central Manchester University
    Hospitals
Manchester, UK

Dongyou Liu
RCPA Quality Assurance Programs
Sydney, New South Wales,
    Australia

Carlos Pérez-Guzmán
General Hospital Third Millennium
Institute of Health
State of Aguascalientes
    Aguascalientes, Mexico

Kesara C. Ratnatunga
Department of Surgery
University of Kelaniya
and
Asiri Surgical Hospital
Colombo, Sri Lanka

Shabbir Susnerwala
Manchester Royal Infirmary
Central Manchester University
    Hospitals
Manchester, UK

1
Introductory Remarks

1.1 Preamble

Tumors or cancers are insidious diseases that result from uncontrolled growth of abnormal cells in one or more parts of the body.* Tumors and cancers have acquired a notorious reputation due not only to their ability to exploit host cellular machineries for their own advantage but also to their potential to cause human misery.

With a rapidly aging world population, widespread oncogenic viruses, and constant environmental pollution and destruction, tumors and cancers are poised to exert an increasingly severe toll on human health and well-being. There is a burgeoning interest from health professionals and the general public in learning about tumor and cancer mechanisms, clinical features, diagnosis, treatment, and prognosis. The following pages in the current volume, as well as those in the sister volumes, represent a concerted effort to satisfy this critical need.

1.2 Tumor mechanisms

The human body is composed of various types of cells that grow, divide, and die in an orderly fashion (so-called apoptosis). However, when some cells in the body change their growth patterns and fail to undergo apoptosis, they often produce a solid and sometimes nonsolid tumor (as in the blood). A tumor is considered benign if it grows but does not spread beyond the immediate area in which it arises. Although most benign tumors are not life-threatening, those found in the brain can be deadly. In addition, some benign tumors are precancerous, with the propensity to become cancer if left untreated. By contrast, a tumor is considered malignant and cancerous if it grows continuously and spreads to surrounding areas and other parts of the body through the blood or lymph system.

A tumor located in its original (primary) site is known as a *primary tumor*. Cancer that spreads from its original (primary) site via the neighboring

---

* The terms *tumor* and *cancer*, along with *neoplasm* and *lesion*, are used interchangeably in colloquial language and publications (see Glossary).

tissue, the bloodstream, or the lymphatic system to another site of the body is called *metastatic cancer* (or *secondary cancer*). Metastatic cancer has the same name and the same type of cancer cells as the primary cancer. For instance, metastatic cancer in the brain that originates from breast cancer is known as *metastatic breast cancer*, not *brain cancer*.

Typically, tumors and cancers form in tissues after the cells undergo genetic mutations that lead to abnormal changes known as *hyperplasia, metaplasia, dysplasia, neoplasia*, and *anaplasia* (see Glossary). Factors contributing to genetic mutations in the cells may be chemical (e.g., cigarette smoking, asbestos, paint, dye, bitumen, mineral oil, nickel, arsenic, aflatoxin, wood dust), physical (e.g., sun, heat, radiation, chronic trauma), viral (e.g., EBV, HBV, HPV, HTLV-1), immunological (e.g., AIDS, transplantation), endocrine (e.g., excessive endogenous or exogenous hormones), or hereditary (e.g., familial inherited disorders).

In essence, tumorigenesis is a cumulative process that demonstrates several notable hallmarks, including (i) sustaining proliferative signaling, (ii) activating local invasion and metastasis, (iii) resisting apoptosis and enabling replicative immortality, (iv) inducing angiogenesis and inflammation, (v) evading immune destruction, (vi) deregulating cellular energetics, and (vii) genome instability and mutation.

1.3 Tumor classification, grading, and staging

Tumors and cancers are usually named for the organs or tissues in which they originate (e.g., brain cancer, breast cancer, lung cancer, lymphoma, skin cancer, etc.). Depending on the type of tissue involved, tumors and cancers are grouped into a number of broad categories: (i) carcinoma (involving the epithelium), (ii) sarcoma (involving soft tissue), (iii) leukemia (involving blood-forming tissue), (iv) lymphoma (involving lymphocytes), (v) myeloma (involving plasma cells), (vi) melanoma (involving melanocytes), (vii) central nervous system cancer (involving the brain or spinal cord), (viii) germ cell tumor (involving cells that give rise to sperm or eggs), (ix) neuroendocrine tumor (involving hormone-releasing cells), and (x) carcinoid tumor (a variant of neuroendocrine tumor found in the intestine).

Tumors of the head and neck include those affecting the ears, eyes, larynx, lips and oral cavity, nasal cavity and paranasal sinus, odontogenic apparatus, pharynx (nasopharynx, oropharynx, and hypopharynx), and salivary glands [1]. Tumors of the cardiovascular and respiratory systems include those affecting the heart, pleura, and lungs [2]. Tumors of the digestive system include those affecting the anus, appendix, bile duct, colon and rectum, esophagus, gallbladder, liver, pancreas, small intestine, and stomach [3].

Under the auspices of the World Health Organization (WHO), the *International Classification of Diseases for Oncology*, 3rd edition (ICD-O-3) has utilized a five-digit system to classify tumors, with the first four digits being morphology code and the fifth digit being behavior code [4]. The fifth-digit behavior codes for neoplasms include the following range: /0 benign; /1 benign or malignant; /2 carcinoma *in situ*; /3 malignant, primary site; /6 malignant, metastatic site; and /9 malignant, primary or metastatic site. For example, meningioma has an IDC-O code of 9530/0 and is a WHO Grade I tumor; atypical meningioma has an IDC-O code of 9539/1 and is a WHO Grade II tumor; and anaplastic (malignant) meningioma has an IDC-O code of 9530/3 and is a WHO Grade III tumor [4].

To further delineate tumors and cancers and assist in their treatment and prognosis, the pathological stages of solid tumor are often determined by using the TNM system of American Joint Commission on Cancer according to the size and extent of the primary tumor (designated *T*, ranging from TX, T0, T1, T2, T3 to T4), the number of nearby lymph nodes involved (*N*, ranging from NX, N0, N1, N2 to N3), and the presence of distant metastasis (*M*, ranging from MX, M0 to M1) [5]. Therefore, the pathological stage of a given tumor is referred to as T1N0MX or T3N1M0 (with numbers after each letter giving more details about the tumor or cancer). However, for convenience, five less-detailed stages (0, I, II, III, and IV), which are based on results of clinical exam and various tests in the absence of findings during surgery, are used clinically to guide the treatment of solid tumors (see Stage, TNM in Glossary) [5].

Another staging system that is more often used by cancer registries than by doctors divides tumors and cancers into five categories: (i) *in situ* (abnormal cells are present but have not spread to nearby tissue); (ii) localized (cancer is limited to the place where it started, with no sign that it has spread); (iii) regional (cancer has spread to nearby lymph nodes, tissues, or organs); (iv) distant (cancer has spread to distant parts of the body); and (v) unknown (there is not enough information to figure out the stage).

1.4 Tumor diagnosis

Because most tumors and cancers tend to induce nonspecific, noncharacteristic clinical signs, a variety of procedures and tests are employed during diagnostic work-up. These involve medical history review of the patient and relatives (for clues to potential risk factors that enhance cancer development), complete physiological examination (for lumps and other abnormalities), imaging techniques, histological evaluation of biopsy and tissue, and laboratory analysis.

The most commonly used imaging techniques include magnetic resonance imaging (MRI), computed tomography (CT) scan, positron emission tomography (PET) scan, and ultrasound. In general, both MRI and CT help reveal the precise location and dimension of tumor mass, but MRI appears superior to CT in terms of image resolution. MRI consists of T1-weighted, T2-weighted, fluid-attenuated inversion recovery (FLAIR), and diffusion-weighted imaging (DWI) sequences. T1-weighted images with or without intravenous contrast gadolinium reveal anatomic details of tumor and surrounding tissue; T2-weighted images highlight tissues with high water concentration (e.g., edema), giving them a white or hyperintense appearance; FLAIR sequence enhances the image of a lesion (e.g., edema). DWI sequence assists in visualization of areas of acute infarction. PET uses a radioactive glucose (sugar) to highlight malignant tumor cells due to their more active uptake of glucose than normal cells.

For histological evaluation, biopsy tissue is stained with hematoxylin and eosin or immunohistochemical dyes, and observed under a microscope. This helps detect cellular abnormalities and verify whether tumor/cancer cells are present at the edge of the material removed (positive margins), or not (negative, not involved, clear, or free margins), or whether they are neither negative nor positive (close margins).

Laboratory analysis of tissue and body fluids (e.g., blood, urine) reveals altered levels of substances in the body, including metabolites, enzymes, proteins, and nucleic acids. The most commonly used methods include biochemical tests, fluorescence *in situ* hybridization, polymerase chain reaction, Southern and Western blot hybridizations, flow cytometry, etc.

1.5 Tumor treatment and prognosis

Standard cancer treatments consist of surgery (for removal of tumor and relieving symptoms associated with tumor), radiotherapy (also called *radiation therapy* or *X-ray therapy*; delivered externally through the skin or internally [brachytherapy] for destroying cancer cells or impeding their growth), chemotherapy (for inhibiting the growth of cancer cells, suppressing the body's hormone production or blocking the effect of the hormone on cancer cells, etc., usually via bloodstream or oral ingestion), and complementary therapies (for enhancing patients' quality of life and improving their well-being). Depending on the circumstances, surgery may be used in combination with radiotherapy and/or chemotherapy to ensure that any cancer cells remaining in the body are eliminated.

The outcomes of tumor and cancer treatment include (i) cure (no traces of cancer remain after treatment and the cancer will never come back); (ii) remission (the signs and symptoms of cancer are reduced; in a complete remission, all signs and symptoms of cancer disappearing for 5 years or more suggest a cure); and (iii) recurrence (a benign or cancerous tumor comes back after surgical removal and adjunctive therapy).

Prognosis (or chance of recovery) for a given tumor is usually dependent on the location, type, grade of the tumor, patient health status and age, etc. Regardless of tumor and cancer types, patients with lower grade lesions generally have a better prognosis than those with higher grade lesions.

1.6 Future perspectives

Tumors and cancers are a biologically complex disease that is expected to surpass heart disease to become the leading cause of human death throughout the world in the coming decades. Despite extensive past research and development efforts, tumors and cancers remain poorly understood and effective cures remain largely elusive.

The completion of the Human Genome Project in 2003 and the establishment of the Cancer Genome Atlas in 2005 offered the promise of a better understanding of the genetic basis of human tumors and cancers and opened new avenues for developing novel diagnostic techniques and effective therapeutic measures.

Nonetheless, a multitude of factors pose continuing challenges for the ultimate conquering of tumors and cancers. These include the inherent biological complexity and heterogeneity of tumors and cancers, the contribution of various genetic and environmental risk factors, the absence of suitable models for human tumors and cancers, and the difficulty in identifying therapeutic compounds that kill or inhibit cancer cells only and not normal cells. Further effort is necessary to help overcome these obstacles and enhance the well-being of cancer sufferers.

Acknowledgments

Credits are due to a group of international oncologists/clinicians, whose expert contributions have greatly enriched this volume.

References

1. Barnes L, Eveson JW, Reichart P, Sidransky D. *Pathology and Genetics of Head and Neck Tumours. World Health Organization Classification of Tumours*. Lyon: IARC Press; 2005.
2. Travis WD, Brambilla E, Burke AP, Marx A, Nicholson AG. *WHO Classification of Tumours of the Lung, Pleura, Thymus and Heart*. Lyon: International Agency for Research on Cancer; 2015.
3. Bosman FT, Carneiro F, Hruban RH, Theise ND. *WHO Classification of Tumours of the Digestive System*. 4th edition. World Health Organization; International Agency for Research on Cancer. Lyon: IARC Press; 2010.
4. Fritz A, Percy C, Jack A, Shanmugaratnam K, Sobin L, Parkin DM, Whelan S. *International Classification of Diseases for Oncology*. 3rd edition. Geneva: World Health Organization; 2000.
5. Edge SB, Byrd DR, Compton CC, Fritz AG, Greene FL, Trotti A, (editors). *AJCC Cancer Staging Manual*. 7th edition. New York: Springer; 2010.

SECTION I
Head and Neck

2
Ear Tumors

2.1 Definition

A variety of tumors can occur in different parts of the ear, including the outer ear (auricle), the middle ear, the inner ear and the temporal bone. Based on their pathological behavior, primary tumors of the ear are differentiated into malignant and benign categories.

Malignant tumors of the ear consist of basal cell carcinoma (BCC), squamous cell carcinoma (SCC), melanoma, adenocarcinoma, acinic cell carcinoma, adenoid cystic carcinoma, osteosarcoma, chondrosarcoma, rhabdomyosarcoma, metastatic carcinoma, lymphoma, malignant neuroma, and malignant paraganglioma [1].

Benign tumors of the ear include neurofibroma, glomus tumor (synonyms: *paraganglioma* and *chemodectoma*), ceruminoma, choristomas, adenoma, schwannoma, chordoma, lipoma, seborrheic keratosis (or *basal cell papilloma*), atheroma (or *sebaceous cyst*), granuloma fissuratum, actinic keratosis, squamous papilloma, sebaceous adenoma, pilomatrixoma, osteoma, exostosis, and keloid [1].

Out of these, BCC arising from the basal layer of the epidermis represents the most common type of ear and temporal bone cancer; SCC originating from keratinocytes in the epidermal layer is the second most common invasive skin cancer (after BCC) of the ear; and glomus tumor (paraganglioma) is a benign tumor that is the most common neoplasm of the middle ear, and the second most common neoplasm of the temporal bone.

In addition, some tumors of the brain (e.g., meningioma, chordoma), head and neck (e.g., parotid malignancy, nasopharyngeal cancer), breast, lung, kidney, and prostate may metastasize to the ear.

2.2 Biology

Structurally, the ear can be divided into the outer ear, the middle ear, the inner ear, and the temporal bone. The *outer ear* (auricle) is composed of cartilage, elastic tissue, and skin that are molded into ridges, hollows, and furrows forming an irregular, shallow funnel with six clearly identified parts: the helix, tragus, lobule, concha (the cup-shaped, deepest depression of the outer ear, leading directly to the external auditory canal), external auditory canal (ear canal or acoustic meatus, which begins at the opening of the concha and extends downward to the eardrum), and eardrum. The *middle ear* is a small cavity comprising three small bones (auditory ossicles), which pass on the vibrations from the eardrum to the inner ear. The *inner ear* contains fluid and fluid-filled cavities (which assist balance) as well as a small spiral tube called the *cochlea*. With many tiny hairlike nerves, the cochlea converts the vibrations from the middle ear into nerve impulses, which are subsequently relayed to the brain. Forming part of the skull surrounding the ear, the *temporal bone* consists of three portions: squamous, tympanic, and petrous. Contained in the squamous portion, the mastoid bone (process) is the lumpy bit behind the ear, with solid outside, honeycomb-like inside, and small cavities with air in between. The mastoid bone also contains the inner ear and the nerves that control the movement of the face and tongue. Thus the temporal bone protects the external auditory canal, middle ear (a space between the squamous and temporal portions laterally and the petrous portion medially of the temporal bone), and inner ear (surrounded by the mastoid bone).

The skin, cartilage, and soft tissue of the ear offer minimal resistance to tumor spread. Tumors in the external auditory meatus can invade posteriorly through the soft tissue into the retroauricular sulcus over the mastoid cortex. Although the bony canal is more resistant to cancer extension, erosion through the posterior bony canal provides access to the mastoid cavity. In addition, tumor growth medially along the external auditory canal can extend through the tympanic membrane and bony tympanic ring into the middle ear. In the middle ear or mastoid, tumors spread easily via the eustachian tube and neurovascular structures beyond the temporal bone to the infratemporal fossa, nasopharynx, or neck.

Tumors that affect the outer ear include BCC, SCC, melanoma, and some benign tumors (e.g., seborrheic keratoses, atheroma, granuloma fissuratum, actinic keratosis, squamous papilloma, sebaceous adenoma, pilomatrixoma, and keloid). Tumors affecting the middle ear are glomus tumor, SCC, schwannoma, middle ear adenoma, choristoma, hemangioma and vascular malformation, Langerhans cell histiocytosis, and primary extracranial meningioma. Acoustic neuroma (vestibular schwannoma) affects the inner ear. Finally, tumors affecting the temporal bone are SCC, glomus tumor, adenocarcinoma, melanoma, rhabdomyosarcoma, osteosarcoma, lymphoma, adenoid cystic carcinoma, and acinic cell carcinoma.

2.3 Epidemiology

Tumors of the ear are rare, with incidence of 1–6 cases per million, accounting for <0.2% of all tumors of the head and neck. A survey of 47 patients with ear tumors revealed a mean age of 54.6 years at diagnosis, and a male-to-female ratio of 3:1 [2]. Further, 62% of the patients had tumors in the middle ear and 38% had tumors in the outer ear. Another study of 27 patients with tumors in the ear and temporal bone identified SCC (74%), BCC (18%), angiosarcoma (4%), and anaplastic carcinoma (4%) [3].

2.4 Pathogenesis

Rick factors for ear tumors include exposure to sun (ultraviolet), radiotherapy, certain industrial chemicals and oils, a history of chronic ear infections (e.g., chronic suppurative otitis media and human papillomavirus for 10 years or longer), a tumor near the ear, cutaneous immunosuppression, transmission through inheritance (e.g., glomus tumor), and genetic mutations (e.g., TP53). Fair skinned people have a greater risk of developing temporal bone cancer.

2.5 Clinical features

Patients with ear tumors may present with pulsatile tinnitus (73%), hearing loss (conductive 49%, mixed 11%, sensorineural 6%), aural pressure/fullness (39%), vertigo/dizziness (16%), otalgia (16%), and bloody otorrhea (6%), earache, headache, ringing sensation in the ear, imbalance, and facial nerve dysfunction (e.g., palsy, twitch, or paralysis) on the side of the affected ear.

Specifically, the most common symptoms of middle ear tumors are blood-stained discharge from the ear, hearing loss, earache and facial nerve dysfunction (eg, palsy, twitch, as seen in facial nerve schwannoma). Those of inner ear tumors are headache, hearing loss, pulsatile tinnitus and dizziness.

2.6 Diagnosis

Patients suspected of ear tumors are examined for ulcers, mass lesions, soft tissue swelling or induration, old scars, otorrhea, facial weakness, and hearing loss. CT and MRI provide further details on the extent of the tumor in relation to surrounding structures. Histological characteristics provide a definitive diagnosis and grading of ear tumors.

BCC usually begins with a scaly area of skin on the ear, which then turns into a slow-growing pearly white lump. The lump (papule or nodule) is reddish-tan to pink in color, with or without central ulceration, and often with telangiectasia (prominent subepidermal vessels). BCC can spread to the inside of the ear. Microscopically, BCC contains basaloid cells with scant cytoplasm and elongated hyperchromatic nuclei, peripheral palisading, peritumoral clefting and mucinous alteration of surrounding stroma, together with mitotic figures and apoptotic bodies. The presence of myxoid stroma and peripheral clefting are most helpful to distinguish BCC from other basaloid tumors. Secondary features include dystrophic calcification, amyloid deposition, or inflammatory reactions with or without partial regression. BCC stains positive for cytokeratin, cytokeratin 5 (or cytokeratin 5/6), nuclear p63, Ber-EP4, and smooth muscle actin (SMA) but negative for epithelial membrane antigen (EMA) [4].

SCC is a plaque with a smooth, irregular, ulcerated, or hyperkeratotic surface. Microscopically, SCC shows full-thickness epidermal replacement by crowded keratinocytes that demonstrate disordered dyspolarity, loss of maturation, and nuclear pleomorphism with hyperchromasia. Apoptotic or dyskeratotic cells as well as typical and atypical mitosis are present at all levels of the epidermis. There is variable loss of the granular layer with surface parakeratosis. Involvement of hair follicles is common. Cutaneous SCC stains positive for cytokeratin, cytokeratin 5 (or cytokeratin 5/6), nuclear p63, and EMA but negative for Ber-EP4 and SMA [5].

Glomus tumor (paraganglioma) is a benign tumor that arises from the paraganglia situated throughout the temporal bone (including on the jugular dome, the promontory of the middle ear, and along the Jacobson and Arnold nerves). Glomus tumor involving the dome of the jugular bulb and spreading to the neck is referred to as *glomus jugulare tumor*, whereas that involving the medial wall along the course of the Jacobson nerve but limited to the ear is referred to as *glomus tympanicum tumor*. Macroscopically, the tumor is a soft, pulsatile, reddish-purple, vascular lobulated mass in the middle ear. On MRI, the tumor is isointense on T1-weighted images and brightly enhanced with gadolinium; and demonstrates increased signal intensity in the solid portions with persistent flow voids in the vascular portions (characteristic "salt-and-pepper" pattern) on T2-weighted images. Histologically, the tumor shows characteristic clusters of chief cells (termed *zellballen* or "cell balls") in a highly vascular stroma. Sustentacular cells and nerve axons, seen in normal paraganglia, rarely appear in paraganglioma.

Immunohistochemically, the tumor is positive for neuron-specific enolase and chromogranin, and occasionally positive for S100 protein, but negative for cytokeratin.

Differential diagnoses of BCC, SCC and glomus tumor of the ear include vestibular schwannoma (or acoustic neuroma, which is benign tumor arising from the schwann cells of the balance nerve located in the internal auditory canal, cerebellopontine angle, or jugular foramen; and which stains positive for S100, but negative for keratin, chromogranin, and synaptophysin), and middle ear adenomas, etc.

Staging of ear tumors is often based on the TNM staging system of the American Joint Commission on Cancer (AJCC), in which T refers to the tumor size and location, N to whether the lymph nodes contain any cancer cells, and M to whether the tumor has metastasized to another area of the body [6]. Another widely used staging system for ear tumors is the University of Pittsburgh TNM staging system. In this classification, a T2 or T3 N1 tumor is considered to be Stage IV, in contrast with the AJCC staging system (sixth edition), in which a T2 or T3 N1 "skin" tumor would be Stage III [7].

2.7 Treatment

The main treatments for ear tumors are surgery, radiotherapy, and chemotherapy. Because some ear tumors (e.g., SCC of the middle ear and mastoid) are fatal without treatment, early diagnosis, surgery, and radiation therapy are crucial to reduce mortality [2,8].

Radical surgery is indicated if the tumor is near the edge of the eardrum. Cancer of the ear canal, glomus tumor (paraganglioma) of the middle ear, and vestibular schwannoma of the balance nerve near the brain often require surgery.If feasible, the tumor together with an area (5 mm all round the cancer) of surrounding tissue completely free of cancer cells (so called a clear margin of tissue) is removed during surgery. The operation to remove the temporal bone containing tumor is called a mastoidectomy or temporal bone resection.

Radiotherapy is advised after tumor excision or in cases where complete resection is not possible. This helps destroy any cell nests of tumor that remain in adjacent areas. Chemotherapy with anticytotoxic drugs (e.g., fluorouracil and cisplatin) helps destroy or inhibit cancer cells.

2.8 Prognosis

In general, patients with lower grades/stages of ear tumors have better prognosis than those with higher grades/stages. Estimated median survivals by stage are Stage I, no deaths; Stage II, >24 months; Stage III, 16 months (range 10–22 months); and Stage IV, 9 months (range 7–11 months).

References

1. Beyea JA, Moberly AC. Squamous cell carcinoma of the temporal bone. *Otolaryngol Clin North Am*. 2015;48(2):281–92.
2. Chao CK, Sheen TS, Shau WY, Ting LL, Hsu MM. Treatment, outcomes, and prognostic factors of ear cancer. *J Formos Med Assoc*. 1999;98(5):314–8.
3. Martinez-Devesa P, Barnes ML, Milford CA. Malignant tumors of the ear and temporal bone: a study of 27 patients and review of their management. *Skull Base*. 2008;18(1):1–8.
4. PathologyOutlines.com website. *Basal cell carcinoma (BCC)*. http://www.pathologyoutlines.com/topic/skintumornonmelanocyticbcc.html. Accessed December 1, 2016.
5. Medscape. *Pathology of squamous cell carcinoma and Bowen disease*. http://emedicine.medscape.com/article/1960631-overview#a7; Accessed December 1, 2016.
6. Edge SB, Byrd DR, Compton CC, et al., (eds.). *AJCC Cancer Staging Manual*. 7th edn. New York: Springer; 2010, pp. 69–78.
7. Arriaga M, Curtin H, Takahashi H, Hirsch BE, Kamerer DB. Staging proposal for external auditory meatus carcinoma based on preoperative clinical examination and computed tomography findings. *Ann Otol Rhinol Laryngol*. 1990;99:714–21.
8. Cristalli G, Manciocco V, Pichi B, et al. Treatment and outcome of advanced external auditory canal and middle ear squamous cell carcinoma. *Craniofac Surg*. 2009;20(3):816–21.

3
Eye Tumors

3.1 Definition

Primary eye tumors (or *ocular tumors*) are commonly divided into two categories: intraocular tumors (starting inside the eyeball) and extraocular tumors (starting outside the eyeball).

Intraocular tumors consist of those affecting children and those affecting adults. Those affecting children include retinal tumor (retinoblastoma), iris and ciliary body lesions (medulloepithelioma), choroidal and retinal pigment epithelium (RPE) lesions (congenital hypertrophy of the RPE, combined hamartoma of the retina and RPE, and congenital melanocytosis), and other benign tumors. Tumors affecting adults include choroidal and RPE lesions (choroidal nevus, and choroidal melanoma), iris and ciliary body lesions (Fuchs adenoma, iris nevus, ciliary body nevus, iris melanoma, and ciliary body melanoma), lymphoma, and metastatic disease to the choroid. Finally, vascular tumors affect both children and adults (retinal capillary hemangioma, retinal cavernous hemangioma, and choroidal hemangioma) [1].

Extraocular tumors include basal or squamous cell carcinoma (BCC or SCC) of the eyelid, rhabdomyosarcoma, conjunctival tumors (SCC, melanoma, Kaposi's sarcoma, lymphoma, epibulbar dermoid, pinguecula, pterygium), and lacrimal gland tumor.

3.2 Biology

As the organ that collects light and sends messages to the brain to form a picture, the eye is composed of three main parts: the eyeball, the orbit, and the adnexal structures.

The *eyeball* (globe) is mostly filled with a jelly-like material called *vitreous humor*, and comprises three main layers: the sclera, the uvea, and the retina. The *sclera* is the tough, white, outer wall of the eyeball. In the front of the eye it is continuous with the cornea, which is clear to let light through. The *uvea* is the middle layer of the eyeball that nourishes the eye. It is where most melanomas of the eye develop. The uvea has three main parts: the iris, the choroid, and the ciliary body. The iris is the colored part of the eye (blue or brown) surrounding the pupil. The choroid is a thin, pigmented layer (consisting of connective tissue and melanocytes) underneath the retina. The ciliary body contains the muscular tissue that produces aqueous humor—a clear fluid—in the front of the eye between the cornea and the lens and also helps the eye focus. The *retina* is a thin-layered structure that lines the eyeball. It contains specialized nerve cells that are sensitive to light as well as blood vessels.

The *orbit* (eye socket) consists of the tissues surrounding the eyeball. These include muscles that move the eyeball in different directions and the nerves attached to the eye. Cancers of these tissues are called *orbital cancers*.

The *adnexal (accessory) structures* include the eyelids and tear glands (e.g., lacrimal gland). Cancers that develop in these tissues are called *adnexal cancers*.

Eye melanoma initiates in pigment-making melanocytes located in the eyeball, conjunctiva (covering the front of the eyeball), and eyelid. Melanoma of the eyeball is often referred to as *uveal* or *choroidal melanoma* because 90% of eyeball melanomas grow in the choroid (sometimes called the uvea, the middle layer of the eyeball) and the remainder in the iris and ciliary body.

Tumors of the eyes are mainly found in the choroid (60%), conjunctiva (9%), ciliary body (7%), retina (7%), orbit (5%), lacrimal gland and duct (2%), cornea (<1%), and unspecified locations (10%). Tumors may also affect areas around the eyeball (e.g., the orbit and adnexa) involving muscle, nerve, and skin. Rhabdomyosarcoma starts in the muscles (the ciliary muscles) that move the eye, usually in children about 6 years of age [2]. Lacrimal gland tumor is a benign or malignant tumor of the glands that produce tears.

3.3 Epidemiology

Tumors of the eye represent <1% of total neoplasm cases. Primary eye tumors affect people at any age, although older people appear to be more vulnerable. About 65% of choroidal melanomas are found in people over 50 years of age. The median age of diagnosis is about 18 months for retinoblastoma and 4–5 years for medulloepithelioma. The cases of eye tumors are evenly distributed between males and females.

3.4 Pathogenesis

Risk factors for eye melanoma include exposure to sunlight and UV radiation, older age (average age at diagnosis is 55), eye color (blue, grey, or green eyes > brown eyes), presence of abnormal moles on the skin (as in dysplastic nevus syndrome), abnormal brown spots on the uvea (as in oculodermal melanocytosis or nevus of Ota), and genetic mutations (*GNA11*, *GNAQ*, and *BAP1* genes; gain or loss of chromosomes 3, 6, and 8). Risk factors for SCC of the eye consist of sun exposure, HIV/HPV infection, and immune therapy (e.g., organ transplantation). Lymphoma of the eye may be related to HIV and *Chlamydophila psittaci* infection, organ transplantation, autoimmune diseases (e.g., rheumatoid arthritis), and older age (50–60 years), whereas ocular Kaposi's sarcoma is associated with HIV infection. Risk factors for retinoblastoma are young age (<5 years) and mutation in the retinoblastoma gene (RB1). Medulloepithelioma also tends to occur in young age (4–5 years).

3.5 Clinical features

People with early stage melanomas of the choroid, ciliary body, and uvea may be asymptomatic, but those with advanced tumors or tumors in certain parts of the eye often display various signs. These include: (i) vision abnormality (blurry vision/double vision, partial or total loss of vision, due to retinal detachment); (ii) floaters (spots or squiggles drifting in the field of vision); (iii) growing dark spot/patch on iris (the colored part of the eye, as in iris melanoma); (iv) change in the size or shape of the pupil (the dark spot in the center of the eye); (v) change in position of the eyeball within its socket; (vi) bulging of the eye (and associated pain); (vii) growing lump on eyelid or in eye; (viii) change in the way the eye moves within the socket; and (ix) redness or swelling in the eye and sensitivity to light [3].

Children with retinoblastoma may present with strabismus (crossed eyes), leukocoria (a white/yellow dot instead of the red eye reflex through the pupil), reduction/loss of vision, and occasional pain and redness in the eye [4]. Medulloepithelioma commonly manifests as a gray-white lesion of the anterior chamber angle or as a diffuse mass causing leukocoria. Clinical features of neovascular glaucoma with a normal posterior segment, iris notching, and an unexplained cyclitic membrane in a child are suggestive of the disease.

3.6 Diagnosis

Diagnosis of eye tumors involves eye examination (with a lighted instrument called ophthalmoscope and a slit lamp, consisting of a microscope/small magnifying lens and a light), ultrasound, CT, MRI, PET, fluorescein and indocyanine green angiography (for ruling out eye problems other than cancer), fine-needle biopsy, bone marrow examination (for lymphoma), cytogenetics (analyzing the number, size, shape, and arrangement of the chromosomes), and gene expression profiling (identifying specific genes, proteins, and other factors unique to the tumor).

Originating from the iris, ciliary body, or choroid, uveal melanoma represents the most common primary intraocular malignant tumor. Macroscopically, choroidal melanoma appears as a dome or mushroom-shaped mass that protrudes into the vitreous, and ultrasound classically shows an elevated solid tumor with low to medium internal reflectivity and significant vascularity. Histologically, melanomas show spindle (longer and tapered at the ends), epithelioid (oval-shaped), and mixed (both spindle and epithelioid) cells; and those containing a higher proportion of epithelioid cells have poorer prognosis [3].

Retinoblastoma typically demonstrates on ultrasound a mass with high reflectivity and intralesional calcium causing shadowing behind the tumor. Histopathologically, retinoblastoma is characterized by basophilic cells with minimal cytoplasm surrounding a lumen in a rosette formation or radially arranged around a central tangle of fibrils in a pseudorosette formation. Necrosis and hemorrhage are often present within the tumors [4].

Medulloepithelioma is a unilateral congenital disease, arising from the epithelium of the medullary tube, particularly the ciliary body. On ultrasound, the tumor displays cystic spaces and a lack of calcification. Histologically, the tumor is composed of epithelium arranged in cords and sheets separated by cystic spaces containing proteinaceous material. In the teratoid form, heterotopic elements including skeletal muscle and cartilage are observed. In the nonteratoid (malignant) form, poorly differentiated neuroblastic cells, increased mitotic activity, sarcomatous areas, and invasion of other ocular tissue are noted.

Stages of eye tumors may be determined by the TNM system of AJCC, which looks at the tumor size (T), involvement of lymph nodes (N) and metastasis (M) to another part of the body, resulting in a number-based score (stage I-IV).

For eye melanoma, a simple staging system based on Collaborative Ocular Melanoma Study (COMS) is often used. By looking at the thickness and width of the tumor, eye melanoma is classified as small (1–3 mm thick, and 5–10 mm across), medium (3.1–8 mm thick, and 10–16 mm across) and large (>8 mm thick, and >16 mm across).

3.7 Treatment

Treatments of eye tumors include surgery, radiotherapy, laser therapy, chemotherapy, and targeted therapy [5–8].

Surgical options include enucleation of the eye (removal of the eyeball, but leaving the muscles and eyelids intact, and subsequent fitting of a prosthetic eye); evisceration (removal of the eye contents, leaving the sclera or the white part of the eye); exenteration (removal of the eye, eyelids, and other orbital contents and subsequent fitting of a prosthesis); iridectomy (removal of the affected piece of the iris, which is the colored part of the eye); choroidectomy (removal of the choroid layer, which is the vascular tissue between the sclera and the retina); iridocyclectomy (removal of the iris plus the ciliary body muscle); iridotrabeculectomy (removal of part of the iris, plus a small piece of the outer part of the eyeball); and eyewall resection (cutting into the eye to remove a tumor).

Radiation therapy enables preservation of the eye structure and saves some of the vision in the eye. It can be carried out by using brachytherapy, external beam radiation therapy, conformal proton beam radiation therapy, or stereotactic radiosurgery via gamma knife or CyberKnife.

Laser therapy applies beams of light to destroy body tissues and is sometimes used to treat intraocular (eye) melanoma but not intraocular lymphoma. It consists of several types. Transpupillary thermotherapy uses infrared light to heat and kill small choroidal melanomas. Laser photocoagulation uses a highly focused, high-energy light beam to burn tissue.

Chemotherapy involves injection of drugs (e.g., methotrexate) into the eye or a vein or oral ingestion to treat tumors that have spread (e.g., intraocular lymphoma).

Targeted therapy includes immunotherapy (using pembrolizumab [Keytruda®] and ipilimumab [Yervoy®] to help the body's own immune system recognize and attack tumor cells) and targeted drugs (targeting the BRAF gene in skin melanomas and CD20 in lymphoma cells with monoclonal antibodies rituximab [Rituxan®] and ibritumomab tiuxetan [Zevalin]) [8].

3.8 Prognosis

People with eye melanoma have a 5-year overall survival rate of 75%, slightly higher (80%) if tumor has not spread and only about 15% if tumor has spread to distant parts of the body. In addition, people with small choroidal melanoma have a 5-year survival rate of 80%, those with medium choroidal melanoma have a rate of 70%, and those with large choroidal melanoma have a rate of 50%.

References

1. Sagoo MS, Mehta H, Swampillai AJ, et al. Primary intraocular lymphoma. *Surv Ophthalmol.* 2014;59(5):503–16.
2. Gündüz K, Yanık Ö. Myths in the diagnosis and management of orbital tumors. *Middle East Afr J Ophthalmol.* 2015;22(4):415–20.
3. Blum ES, Yang J, Komatsubara KM, et al. Clinical management of uveal and conjunctival melanoma. *Oncology (Williston Park).* 2016;30(1):29–32, 34–43, 48.
4. Faranoush M, Hedayati Asl AA, Mehrvar A, et al. Consequences of delayed diagnosis in treatment of retinoblastoma. *Iran J Pediatr.* 2014;24(4):381–6.
5. Buder K, Gesierich A, Gelbrich G, Goebeler M. Systemic treatment of metastatic uveal melanoma: Review of literature and future perspectives. *Cancer Med.* 2013;2(5):674–86.
6. Sikuade MJ, Salvi S, Rundle PA, Errington DG, Kacperek A, Rennie IG. Outcomes of treatment with stereotactic radiosurgery or proton beam therapy for choroidal melanoma. *Eye (Lond).* 2015;29(9):1194–8.
7. Yousef YA, Alkilany M. Characterization, treatment, and outcome of uveal melanoma in the first two years of life. *Hematol Oncol Stem Cell Ther.* 2015;8(1):1–5.
8. Ostrowski RA, Bussey MR, Shayesteh Y, et al. Rituximab in the treatment of thyroid eye disease: A review. *Neuroophthalmology.* 2015;39(3):109–15.

4
Laryngeal Cancer

4.1 Definition

The larynx is affected by a variety of tumors, chief among which are squamous cell carcinoma (SCC), adenocarcinoma, neuroendocrine tumor, and sarcoma.

Laryngeal SCC develops in the thin, flat, scalelike cells that line much of the larynx and accounts for 95% of laryngeal cancer reported. Characterized by squamous differentiation, SCC may be divided into keratinizing and non-keratinizing subtypes and well, moderately, and poorly differentiated grades. Variants of laryngeal SCC include verrucous carcinoma, basaloid SCC, papillary SCC, spindle SCC (or spindle cell carcinoma), acantholytic SCC, and adenosquamous carcinoma [1].

Laryngeal adenocarcinoma begins in the glandular cells present in the wall of the larynx that make mucus and is a relatively uncommon neoplasm in this location. Laryngeal neuroendocrine tumor (carcinoid) and sarcoma are even rarer malignancies of the larynx [2].

4.2 Biology

The larynx (or the voice box) is a 5-cm long structure located in the inferior portion of the pharynx and superior to the trachea. The larynx consists of three anatomical regions: the supraglottis (including epiglottis, false vocal cords, ventricles, arytenoids, and aryepiglottic folds); the glottis (including true vocal cords, anterior and posterior commissures); and the subglottis (starting about 1 cm below the true vocal cords and extending to the end of the cricoid cartilage or the first tracheal ring).

Histologically, the true vocal folds are composed of a layer of stratified squamous epithelium overlying the lamina propria, a gel-filled space consisting of superficial, middle, and deep layers. The middle and deep layers make up the vocal ligament, and the superficial layer is the gel layer. There are multiple connective tissue barriers within the larynx.

The primary functions of the larynx are to protect the lower airway during swallowing, to breathe through the rima glottidis, and to produce sound (phonation) through vibration.

Laryngeal SCC often evolves in a background of mucosal squamous dysplasia or carcinoma *in situ*. Based on tumor location, laryngeal SCC may be separated into glottic, supraglottic, subglottic, and transglottic SCC. Glottic SCC arises from true vocal cord, anterior and posterior commissure or vocal processes of arytenoid cartilage; it accounts for 60% of all laryngeal cases. Its invasion is usually limited to tissue superficial to the conus elasticus, vocal ligament, and thyroglottic ligament. It has a high cure rate due to the lack of lymphatics in the true vocal cords but may spread to the opposite cord. Supraglottic SCC arises above the true vocal cords from the epiglottis, ventricle, angle between the epiglottis and ventricle, aryepiglottic fold, arytenoid body, and false vocal cords; it accounts for 30% of all cases. It does not usually invade the thyroid cartilage, and 40% have nodal metastases. Subglottic SCC (also called *infraglottic SCC*) arises from regions inferior to the vocal cords to the lower border of the cricoid cartilage and accounts for <5% of all cases. It often spreads laterally to the cricoid cartilage and destroys the interthyrocricoid membrane with invasion of the prelaryngeal wall and thyroid gland. It may be involved in cervical nodal metastases (15%) and paratracheal nodal metastases (50%). Transglottic SCC crosses the ventricle vertically, is associated with the fixed vocal cord, and accounts for <5% of all cases. It has the highest incidence of nodal metastases (52%) [3].

4.3 Epidemiology

Laryngeal cancer (sometimes called *throat cancer* despite the fact that the throat includes the larynx and pharynx) represents 3% of head and neck cancer cases, with global incidence of 3.2 cases per 100,000 and a death rate of 1.1 per 100,000 per annum.

In Spain, Italy, France, Brazil, India, and the Afro-Caribbean populations in parts of the United States, the incidence is high (>10 per 100,000), whereas in Japan, Norway, and Sweden, the incidence is low (2 per 100,000). Laryngeal cancer tends to affect men of 55–65 years old and shows a male predilection (5.8 per 100,000 in men vs. 1.2 per 100,000 in women).

4.4 Pathogenesis

Risk factors for laryngeal cancer include cigarette smoking, excess alcohol consumption, aging, vitamin A deficiency, exposure to asbestos, poor dental hygiene, human papillomavirus (HPV) infection, and presence of squamous cell cancers in the upper aerodigestive tract.

Laryngeal cancer is linked to chromosomal gains in 7q35 and 8q24 and losses in 1p21, 2q21, 17q12, and 3p22, in addition to promoter mutations (e.g., C228T and C250T) in the telomerase reverse transcriptase (*TERT*) gene, *CYP1B1* gene polymorphism, and overexpression of *P53*, *EGFR*, *CCNA2*, *CCNB1*, *CCNB2*, and *CDK1* [4].

4.5 Clinical features

The symptoms of laryngeal cancer range from hoarse voice, lump or swelling in the neck, pain when swallowing, sore throat, persistent cough, stridor (a high-pitched wheezing sound indicative of a narrowed or obstructed airway), bad breath, earache, difficulty with breathing to weight loss.

Specifically, glottic cancer often affects the voice at earlier stages; supraglottic cancer may induce sore throat, trouble swallowing (dysphagia), earache, change in voice quality, or enlarged neck nodes. Early vocal cord cancer causes hoarseness. Subglottic cancer commonly involves the vocal cords and leads to airway obstruction at an earlier stage.

4.6 Diagnosis

Diagnostic procedures for laryngeal cancer include physical exam, medical history review, imaging (CT, PET, MRI), biopsy, histopthological assessment and molecular testing.

Laryngeal SCC is a pink to gray ulcerated mass, with vocal cord lesions being keratotic. Histologically, the tumor contains islands, tongues, and clusters of atypical cells invading the laryngeal stroma; it appears well, moderate, or poorly differentiated, based on degree of keratinization, pearl formation, intercellular bridges and mitotic activity, with a smaller tumor being better differentiated. The tumor stains positive for AE1, AE3, and p53 (50%) [3].

Among laryngeal SCC variants, verrucous SCC is a large, locally destructive, white-tan exophytic tumor of up to 10 cm fixed to normal structures without metastasizing to nearby lymph nodes. Histologically, it is an invasive tumor with well-differentiated squamous epithelium that lacks features of SCC but shows uniform cells without atypia or mitotic figures, marked surface keratinization (church-spire keratosis), broad pegs with pushing but not an infiltrative margin, and prominent lymphoplasmacytic and histiocytic infiltrate. SCC represents 1%–4% of laryngeal cancer and may coexist with conventional SCC [3,5].

Spindle SCC (also called *sarcomatoid carcinoma*, *carcinosarcoma*) is a polypoid (99%) or endophytic mass of 2 cm containing a pleomorphic or storiform spindle cell component with foci of benign or malignant cartilage or bone (typical of *in situ* or invasive SCC) and frequent mitotic activity. The tumor stains positive for vimentin, 34betaE12, and AE1-AE3; variable smooth muscle actin; but negative for CAM 5.2. Spindle SCC is uncommon in the larynx (71% are glottic; 59% are T1) and with nodal metastases may have epithelial or stromal patterns or both; it is more aggressive than conventional SCC [3,5].

Basaloid SCC is a firm to hard, tan-white mass of up to 6 cm, with central necrosis and neck metastases. Microscopically, the tumor shows nests and lobules of small basaloid cells with minimal cytoplasm, hyperchromatic nuclei, comedonecrosis, prominent hyalinization and peripheral palisading, small cystic spaces, and mitotic activity. The tumor stains positive for 34betaE12 (100%), AE1-AE3, CAM5.2, epithelial membrane antigen, carcinoembryonic antigen (53%), S100 (39%), neuron-specific enolase (weak, 75%), periodic acid-Schiff (PAS), and Alcian Blue (material within cystic spaces) but negative for synaptophysin, chromogranin, muscle-specific actin, and glial fibrillary acidic protein (GFAP) [3,5].

Papillary SCC is a solitary, exophytic or papillary mass of 2 mm to 4 cm, with malignant squamous epithelium and limited surface keratinization. Microscopically, the tumor shows fingerlike projections with fibrovascular cores or broad-based bulbous growth with rounded projections and limited fibrovascular cores. The tumor is associated with HPV and has relatively good prognosis (usually T2) [3].

To evaluate potential prognostic indicators of laryngeal SCC, DNA ploidy, proliferative activity (e.g., mitotic index), and oncogene amplification analyses may be utilized.

The stages of laryngeal cancer range from 0, I, II, III, to IV. Stage 0 tumor has not invaded tissue beyond the throat. Stage I tumor is less than 7 cm and limited to the throat. Stage II tumor is slightly greater than 7 cm but still limited to larynx. Stage III tumor has grown and spread to nearby tissues and organs. Stage IV tumor has spread to the lymph nodes or distant organs.

4.7 Treatment

Treatment for laryngeal cancer involves surgery (used in 55% of cases), radiotherapy (70%), or chemotherapy (10%, such as cetuximab or Erbitux), alone or in combination. The ultimate goal of laryngeal cancer treatment is to preserve the voice as much as possible, with approaches ranging from removing a small tumor on the vocal cord(s) by a CO_2 laser; radiotherapy (via intensity-modulated radiotherapy, 3D-conformal radiation therapy, brachytherapy, or proton therapy) for large glottic and supraglottic tumors; laryngectomy (removing a portion or all of the voice box) for residual tumor after radiotherapy and recurrence; debulking and tracheotomy for tumor causing respiration; to laryngectomy and postoperative radiotherapy for very advanced tumor [6].

Over 95% of patients with laryngeal SCC are treatable. Laryngeal SCC *in situ* may be treated by mucosal stripping or superficial laser excision, together with radiation therapy if necessary. Early stage laryngeal SCC can be treated with single-modality therapy, either surgery or radiotherapy, with a 5-year local control of 85%–95%. More advanced laryngeal SCC disease (Stages III and IV) requires a multimodality treatment with either chemoradiation or surgery and radiotherapy [6,7]. Indeed, chemoradiation is often preferred for organ preservation in selected patients. However, laryngectomy and postoperative radiotherapy are necessary for patients who fail primary chemoradiation [7].

Cetuximab is a monoclonal antibody which binds specifically to epidermal growth factor receptors (EGFRs) and prevents the cancer cells from growing and dividing.

Other chemotherapeutic agents for laryngeal cancer inlcude cisplatin, 5-fluorouracil and taxane [7,8]. In Europe, induction chemotherapy with taxane, cisplatin, and fluorouracil (TPF), followed by radiotherapy is favored; whereas in North America, concomitant cisplatin and standard fractionation radiotherapy (CCR) is preferred [8].

4.8 Prognosis

Laryngeal cancer has a 5-year survival rate of about 60%. For localized, regional, distant, and unstaged laryngeal cancer, the 5-year survival rates are 76.3%, 44.5%, 35.1%, and 54.6%, respectively. Five-year survival for patients with extremely advanced local and regional disease is <5%.

Five-year disease-free survival for patients with glottic cancer is 85%–90% for Stages I–II, 75% for Stage III, and 45%–50% for Stage IV. Five-year disease-free survival for patients with supraglottic cancer is 80% for Stages I–II, 70% for Stage III, and 40% for Stage IV (as supraglottic cancer shows an increased incidence of nodal metastases in comparison with glottic cancer).

References

1. Barnes L, Eveson JW, Reichart P, et al. *Pathology and Genetics of Head and Neck Tumours. World Health Organization Classification of Tumours.* Lyon: IARC Press; 2005.
2. Ferlito A, Silver CE, Bradford CR, et al. Neuroendocrine neoplasms of the larynx: An overview. *Head Neck.* 2009;31(12):1634–46.
3. Pathologyoutlines.com website. Larynx and hypopharynx. http://www.pathologyoutlines.com/larynx.html; accessed December 10, 2016.
4. Qu Y, Dang S, Wu K, et al. TERT promoter mutations predict worse survival in laryngeal cancer patients. *Int J Cancer.* 2014;135(4):1008–10.
5. Steuer CE, El-Deiry M, Parks JR, et al. An update on larynx cancer. *CA Cancer J Clin.* 2017;67(1):31–50.
6. PDQ Adult Treatment Editorial Board. *Laryngeal Cancer Treatment (PDQ®): Health Professional Version.* PDQ Cancer Information Summaries. Bethesda, MD: National Cancer Institute (US); 2002–2016.
7. Vainshtein JM, Wu VF, Spector ME, Bradford CR, Wolf GT, Worden FP. Chemoselection: a paradigm for optimization of organ preservation in locally advanced larynx cancer. *Expert Rev Anticancer Ther.* 2013;13(9):1053–64.
8. Forastiere AA, Weber RS, Trotti A. Organ preservation for advanced larynx cancer: Issues and outcomes. *J Clin Oncol.* 2015;33(29):3262–8.

5
Nasal Cavity and Paranasal Sinus Cancer

5.1 Definition

Tumors commonly affecting the mucus-producing tissue (mucosa) of the nasal cavity and paranasal sinuses include squamous cell carcinoma (SCC), malignant lymphoma, malignant melanoma, esthesioneuroblastoma, and sarcomas, etc [1].

In addition, minor salivary gland tumor, and some noncancerous growths (e.g., nasal polyps and papillomas) may occur in the nasal cavity and paranasal sinuses and cause problems.

5.2 Biology

The nasal cavity is the space located behind the nose, which runs along the roof of the mouth (i.e., the palate that separates the nose from the mouth) and then turns downward to the throat.

The paranasal sinuses are air-filled areas surrounding the nasal cavity. They consist of (i) maxillary sinuses (located in the cheeks, below the eyes on either side of the nose), (ii) ethmoid sinuses (located on the bridge of the nose, between the eyes), (iii) frontal sinuses (located above the inner eye and eyebrow area), and (iv) sphenoid sinuses (located deep behind the nose/the ethmoids, between the eyes).

The main functions of the nasal cavity and paranasal sinuses are to filter, warm, and moisten the air during breathing, to provide voice resonance, to lighten the weight of the skull, and to form a bony framework for the face and eyes.

Lining the surface of the nasal cavity and the paranasal sinuses, the mucus-producing mucosa comprises a number of cell types, including squamous epithelial cells (flat cells lining the sinuses), glandular cells (e.g., minor salivary gland cells, which produce mucus and other fluids), nerve cells (which are involved in sensation and the sense of smell in the nose), lymphocytes (which are part of the immune system), blood vessel cells, bone and cartilage cells, and other supporting cells.

The varied nature of nasal cavity and paranasal sinus cancer reflects the diversity of cell types in these areas. Paranasal sinus cancer is often found in the maxillary sinus, ethmoid sinuses, nasal vestibule, and nasal cavity and rarely in the sphenoid and frontal sinuses.

SCC originates from squamous cells and represents the most common type (>50%) of nasal cavity and paranasal sinus cancer. Malignant lymphoma (e.g., T-cell/natural killer cell nasal-type lymphoma, formerly known as *lethal midline granuloma*) derives from lymphocytes of the lymph system in the lining (mucosa) of the nasal cavity and paranasal sinuses. It accounts for 5% of cancer in these areas. Malignant melanoma arises from melanocytes that line the nasal cavity and sinuses or other areas inside the body as well as the skin; it accounts for about 1% of tumors found in the nasal cavity and paranasal sinuses [2,3]. Esthesioneuroblastoma (or olfactory neuroblastoma, formerly *esthesioneuroepithelioma)* starts from the olfactory nerves near the cribriform plate (a bone located deep in the skull between the eyes and the sinuses) on the roof of the nasal cavity. It is similar to neuroendocrine cancer in appearance and may be confused with undifferentiated carcinoma or undifferentiated lymphoma [4]. Sarcomas (e.g., chondrosarcoma, osteosarcoma, Ewing sarcoma, and most soft tissue sarcomas) are tumors of the muscle, bone, cartilage, and fibrous cells that may start anywhere in the body, including the nasal cavity and paranasal sinuses.

Minor salivary gland tumors (including adenocarcinoma, adenoid cystic carcinoma and mucoepidermoid cancer) evolve from minor salivary gland cells, and represent the second most common type (accounting for 10–15%) of nasal cavity and paranasal sinus cancers.

5.3 Epidemiology

Nasal cavity and paranasal sinus cancer is relatively uncommon and tends to affect people of 45–85 years of age, with a male-to-female ratio of 2:1.

Esthesioneuroblastoma (or *olfactory neuroblastoma*) shows a bimodal age distribution with one peak in young adults (approximately the second decade) and another peak in the fifth to sixth decades, with no obvious gender predilection.

5.4 Pathogenesis

Risk factors for nasal cavity and paranasal sinus cancer include tobacco smoking, chewing, and snuff; heavy alcohol use; repeated exposure to inhaled substances (e.g., dusts from wood [carpentry], textiles [textile factory], leather [shoemaking], flour [baking and flour milling], nickel and chromium [mining], mustard gas, asbestos, and radium); and human papillomavirus (HPV) infection.

Inhaled dusts and cigarette smoke are capable of activating macrophages, neutrophils, and T lymphocytes for production of proteases and reactive oxygen species (ROS), which contribute to DNA damage. Infection with HPV can also induce the ataxia telangiectasia mutated (ATM)-dependent DNA damage. The resulting DNA damages in the cells of the nasal cavity and sinuses may switch on oncogenes (which promote cell division) or switch off tumor suppressor genes (which slow down cell division or cause cells to die at the right time), leading to the development of cancers.

5.5 Clinical features

Patients with nasal cavity or paranasal sinus cancer may display nasal obstruction, persistent nasal congestion, and stuffiness (so-called sinus congestion); chronic sinus infections that do not respond to antibiotic treatment; postnasal drip; nosebleeds; pus draining from the nose; frequent headaches; pain in the sinus region, face, eyes, or ears; constant watery eyes; bulging of one of the eyes or vision loss; decreased sense of smell; pain (or numbness) in the teeth; loosening of teeth; difficulty opening the mouth; lump on the face, nose, or inside the mouth; lump in the neck; fatigue; and unexplained weight loss.

Clinical presentation of esthesioneuroblastoma (or *olfactory neuroblastoma*) is usually secondary to nasal stuffiness and rhinorrhoea or epistaxis. Large tumor extending into the intracranial compartment (25%–30% at diagnosis) may cause anosmia. Lethal midline granuloma is associated with rhinorrhea, epistaxis, nasal stuffiness, obstruction, and pain.

Sometimes people with nasal cavity or paranasal sinus cancer do not show any of these symptoms, and nasal cavity or paranasal sinus cancer is only discovered after investigation of inflammatory disease of the sinuses (e.g., sinusitis).

5.6 Diagnosis

Constituting >50% of nasal cavity and paranasal sinus cancers, squamous cell carcinoma (SCC) also commonly occurs in pharyngeal regions, and its diagnosis is detailed in Chapter 8. Minor salivary gland tumors make up

10–15% of nasal cavity and paranasal sinus cancers, and their diagnosis is described in Chapter 9.

Arising from the basal layer of the olfactory epithelium in the superior recess of the nasal cavity, *esthesioneuroblastoma* (or *olfactory neuroblastoma*) is a multilobulated, pink-gray, soft tissue mass involving the anterior and middle ethmoid air cells on one side and extending through the cribriform plate into the anterior cranial fossa, leading to obstruction of the ostia of paranasal sinuses and opacification of the sinus with secretions [5].

On CT, the tumor shows soft tissue attenuation, with relatively homogeneous enhancement and occasional focal calcifications. On MRI, the tumor presents with heterogeneous intermediate signal on T1, heterogeneous intermediate signal on T2, and variable enhancement (usually moderate to intense) on T1 C+ (Gd). On angiography, the tumor blushes with arteriovenous shunting and persistent opacification [5].

As a presentation of nasal type NK/T cell lymphoma, *midline granuloma* (or *lethal midline granuloma*) is a rare midfacial necrotizing disease characterized by destruction and mutilation of the nose and other structures of the respiratory passages (e.g., sinuses and nearby tissues). Linked to natural killer/T-cell non-Hodgkin's lymphoma and Wegener's granulomatosis, the tumor causes thickening in the underlying mucosa, perforation of the nasal septum and palate, erosion of the antral bone, or ulceration of the skin overlying the nose or antrum [4].

Microscopically, the tumor shows polymorphic infiltrate of lymphocytes (small to large), plasma cells, neutrophils, and scattered atypical lymphoid cells with perinuclear clearing; frequent angiolymphatic invasion and necrosis; epitheliotropism; and frequent histiocytes with erythrophagocytosis. The tumor stains positive for CD2, CD3 (cytoplasmic), EBER-ISH (100%), CD56 (65%), p53 (50%), CD45RO, CD43, TIA1, granzyme B, perforin (35%–100%), CD8 (20%), and EBV LMP1 (48%) but negative for CD3 (nuclear), CD16 (positive on histiocytes but not tumor cells), CD57, and CD79a [4,6].

Based on their location in the nasal cavity and paranasal sinuses, status of spread, and effect on other parts of the body, nasal cavity or paranasal sinus cancers (particularly cancers of the maxillary sinus, the nasal cavity, and the ethmoid sinus) may be differentiated into five stages (0, I, II, III, IV) according to the TNM system of the American Joint Committee on Cancer.

Specifically, stage 0 is a very early cancer (Tis) with no spread to lymph nodes (N0) or distant metastasis (M0); stage I is a noninvasive cancer (T1/N0/M0); stage II is an invasive cancer (T2/N0/M0); stage III includes invasive cancer (T3/N0/M0), as well as invasive cancer (T1–T3) that has spread to regional lymph nodes (N1) but shows no sign of metastasis (M0). Stage IVA is an invasive cancer (T4a/N0-N2/M0; any T/N2;M0); stage IVB is an invasive cancer (T4b/any N/M0; any T/N3/M0); and stage IVc is an invasive tumor (any T/any N/M1).

5.7 Treatment

Treatment options for nasal cavity and paranasal sinus cancer include surgery, radiotherapy, chemotherapy, targeted therapy, and palliative treatment [7].

By removing the entire tumor and a rim of surrounding normal tissue, surgery represents an essential part of treatment for nasal cavity and paranasal sinus cancer, especially in patients who fail to respond to radiotherapy.

Radiotherapy may be used alone or after surgery. However, radiotherapy has some side effects, ranging from skin problems (sunburn-like effect), nausea, loss of appetite, feeling tired or weak, mouth/throat pain and sores in the mouth (mucositis), trouble swallowing, hearing loss, hoarseness, problems with taste, to tooth decay, bone pain, and damage.

Chemotherapy (with carboplatin, cisplatin, 5-fluorouracil, docetaxel [Taxotere], paclitaxel [Taxol®], bleomycin, cyclophosphamide [Cytoxan®], vinblastine, and methotrexate) may be utilized along with surgery and/or radiation for more advanced, nonspreading cases, whereas chemotherapy alone is applied to cases with obvious spreading. Notable side effects of chemotherapy consist of nausea and vomiting, loss of appetite, loss of hair, mouth sores, diarrhea, and low blood counts.

Targeted therapy for nasal cavity and paranasal sinus cancer relies on the use of cetuximab (Erbitux®), which may be combined with radiotherapy for some earlier stage cancers. For more advanced cancers, cetuximab may be combined with cisplatin.

Palliative treatment can help ease symptoms from the main cancer treatment itself and maintain quality of life for as long as possible [8].

5.8 Prognosis

Cure rates for patients with advanced tumors in the nasal cavity and paranasal sinuses are generally poor (≤50%), and 20%–40% of patients with metastases do not respond to treatment. The 5-year survival rate (which refs to the percentage of patients who live *at least* 5 years after their cancer is diagnosed) for nasal cavity and paranasal sinus cancer is stage-dependent: 63% for Stage I, 61% for Stage II, 50% for Stage III, and 35% for Stage IV.

References

1. Michel G, Joubert M, Delemazure AS, Espitalier F, Durand N, Malard O. Adenoid cystic carcinoma of the paranasal sinuses: Retrospective series and review of the literature. *Eur Ann Otorhinolaryngol Head Neck Dis.* 2013;130(5):257–62.
2. Mihajlovic M, Vlajkovic S, Jovanovic P, Stefanovic V. Primary mucosal melanomas: A comprehensive review. *Int J Clin Exp Pathol.* 2012;5(8):739–53.
3. Gilain L, Houette A, Montalban A, Mom T, Saroul N. Mucosal melanoma of the nasal cavity and paranasal sinuses. *Eur Ann Otorhinolaryngol Head Neck Dis.* 2014;131(6):365–9.
4. Mallya V, Singh A, Pahwa M. Lethal midline granuloma. *Indian Dermatol Online J.* 2013;4(1):37–9.
5. Radiopedia.org. *Olfactory Neuroblastoma.* https://radiopaedia.org/articles/olfactory-neuroblastoma; accessed December 15, 2016.
6. Pathologyoutlines.com. *NK/T Cell Lymphoma, Nasal Type.* http://www.pathologyoutlines.com/topic/nasalNKlymphoma.html; accessed December 15, 2016.
7. PDQ Adult Treatment Editorial Board. *Paranasal Sinus and Nasal Cavity Cancer Treatment (PDQ®): Health Professional Version.* PDQ Cancer Information Summaries. Bethesda, MD: National Cancer Institute (US); 2002–2015.
8. Jégoux F, Métreau A, Louvel G, Bedfert C. Paranasal sinus cancer. *Eur Ann Otorhinolaryngol Head Neck Dis.* 2013;130(6):327–35.

6
Odontogenic Tumors

6.1 Definition

Odontogenic (tooth-forming) apparatus and its remnants (epithelial and mesenchymal) are affected by a large number of tumors, allied lesions, and cysts [1,2].

Odontogenic tumors comprise a heterogeneous group of neoplasms that can be separated into malignant and benign categories.

Malignant odontogenic tumors include odontogenic carcinomas (e.g., metastasizing malignant ameloblastoma, ameloblastic carcinoma, primary intraosseous carcinoma NOS [or primary intraosseous squamous cell carcinoma, PIOSCC], clear cell odontogenic carcinoma, and ghost cell odontogenic carcinoma) and odontogenic sarcomas (e.g., ameloblastic fibrosarcoma and odontogenic sarcoma NOS). These tumors are serious but fortunately are rare [3].

Benign odontogenic tumors consist of epithelial odontogenic tumors (e.g., ameloblastoma, unicystic-type ameloblastoma, squamous odontogenic tumor, calcifying epithelial odontogenic tumor, and adenomatoid odontogenic tumor), mixed odontogenic tumors (e.g., ameloblastic fibroma, odontoma, and developing odontoma), mesenchymal odontogenic tumors (e.g., odontogenic fibroma, granular cell odontogenic tumor, odontogenic myxoma/myxofibroma, cementoblastoma, and cemento-ossifying fibroma), and peripheral odontogenic tumors. All together, these tumors account for >95% of odontogenic tumors diagnosed [4].

Odontogenic allied lesions (or bone-related lesions) encompass psammomatoid ossifying fibroma, trabecular ossifying fibroma, fibrous dysplasia, cemento-osseous dysplasia, central giant cell granuloma, cherubism, aneurysmal bone cyst, simple bone cyst, and melanotic neuroectodermal tumor of infancy [4].

Odontogenic cysts comprise cysts of inflammatory origin (e.g., radicular cyst, and inflammatory collateral [paradental] cyst) and cysts of unknown origin (dentigerous cyst, odontogenic keratocyst, lateral periodontal cyst, gingival cyst, glandular odontogenic cyst, calcifying odontogenic cyst, and orthokeratinized odontogenic cyst) [4].

6.2 Biology

Situated in the alveolar processes (which protrude from the maxilla and mandible) and held fast in sockets by periodontal ligament, teeth are made up of enamel, dentin, cementum, and pulp. Enamel mostly contains calcium phosphate and forms the hardest, white outer part (crown) of the tooth. Located beneath enamel and surrounding the pulp chamber, dentin is a calcified matrix consisting of living cells (which secrete a hard mineral substance) and neurovascular elements. Being an extension of dentin and containing similar mineralization and connective tissue, cementum covers the root canal (extending from the pulp chamber) and binds the roots of the teeth firmly to the gums and jawbone. Pulp is the softer, living inner structure of teeth with blood vessels and nerves running through it.

Covering the alveolar processes of the maxilla and mandible and finishing at the neck of each tooth, the gingiva (gums) is a mucosal tissue lined with epithelium. The gingival epithelium is divided into the oral, sulcular, and junctional sections. The oral epithelium comprises stratified squamous keratinizing epithelium, which covers the oral and vestibular gingival surfaces. The sulcular epithelium is continuous with the oral epithelium and lines the gingival sulcus. The junctional epithelium lines the dentoepithelial junction at the bottom of the gingival sulcus.

Odontogenic tumors, lesions, and cysts mostly arise from epithelial, ectomesenchymal, and/or mesenchymal elements of the tooth-forming apparatus (e.g., the jaw bones or the gingival mucosa) during or after odontogenesis. The mandible (about 70%) appears to be the most favored site (involving the body, angle, and vertical ramus) for all odontogenic tumors (especially ameloblastoma), and the maxilla (about 30%) is the next most frequent site (mostly in the posterior region) for ameloblastic carcinoma, primary intraosseous squamous cell carcinoma, and clear cell odontogenic carcinoma.

6.3 Epidemiology

Odontogenic tumors, allied lesions, and cysts account for <1% of all oral and maxillofacial clinical cases. The age of patients with odontogenic tumors ranges from 6 to 84 years, with a mean age of 29–34 years. Although ameloblastoma is evenly distributed in both sexes, PIOSCC shows a male-to-female ratio of 2:1.

A recent survey indicated that ameloblastoma (30%), keratocystic odontogenic tumor (26%), odontoma (16%), and calcifying epithelial odontogenic tumor (11%) represent the most common odontogenic tumors, of which benign and malignant tumors account for 94% and 6%, respectively.

6.4 Pathogenesis

Odontogenic tumors are linked to loss of heterozygosity at chromosomes 3p, 9q22.3–q31 (PTCH), 11p, 11q, and 17p13.1 (p53 and CHRNB1). Further, activating *BRAF* V600E mutation is often detected in patients with mandibular ameloblastoma (average age of 34.5 years), whereas *BRAF* wild type is found in patients with maxillary ameloblastoma (average age of 53.6 years). Several genes (*KRAS, NRAS,* and *HRAS*) related to the MAPK pathway are implicated in the pathogenesis of ameloblastoma [5,6].

6.5 Clinical features

Most odontogenic tumors occur intraosseously within the maxillofacial skeleton, while others take place extraosseously in the tooth-bearing mucosa. Clinically, nearly all malignant odontogenic tumors present with dull pain, rapidly developing swelling, paresthesia, loose teeth, and destruction of the buccal and lingual cortical plates of the jawbones. In contrast, benign odontogenic tumors tend to have slow expansive growth with no or slight pain.

6.6 Diagnosis

Diagnosis of odontogenic tumors, lesions and cysts is largely based on imaging and histopathologic features. Among *malignant odontogenic tumors*, ameloblastic carcinoma shows sheets, islands, or trabecular epithelium of round, spindled to tall columnar cells, with overtly malignant cytologic features (i.e., nuclear pleomorphism, mitotic activity of two mitoses in a high power field, focal necrosis, and nuclear hyperchromasia). Ameloblastic carcinoma with a limited atypia or intermediate grade cytologic features is designated as "atypical ameloblastoma." SOX2 may be used in conjunction with Ki-67 for differentiation of ameloblastic carcinoma from ameloblastoma and atypical ameloblastoma.

Primary intraosseous carcinoma represents one of the three subtypes within squamous cell odontogenic carcinoma (the others being carcinoma arising from the epithelial lining of an odontogenic cyst and carcinoma arising from benign epithelial odontogenic tumor like keratocystic odontogenic tumor [KCOT]). Histologically, this tumor exhibits sheets or islands of neoplastic squamous epithelium and moderate differentiation without prominent keratinization. PIOSCC derived from odontogenic cysts may display a microscopic transition area from benign cystic epithelial lining to invasive squamous cell carcinoma and adenocarcinoma (SCCA), with no carcinomatous changes in the overlying epithelium and the adjacent structures. PIOSCC derived from KCOT typically contains a keratinizing well-differentiated squamous cell carcinoma in conjunction with KCOT.

Clear cell odontogenic carcinoma is a locally aggressive low grade malignancy presenting as an ill-defined lucency of the jaw (mandible), with the capacity for invasion of medullary bone, nerves, lymphatics, and regional lymph node as well as distant metastases (pulmonary, bone). Histologically, the tumor displays sheets, cords, or nests of polygonal cells separated by a hyalinized to fibrous stroma. The tumor is positive for cytokeratins (AE1/AE3, CK19), p63, and epithelial membrane antigen but negative for vimentin, S100 protein, desmin, smooth muscle actin, human melanoma antigen (HMB-45), α1-antichymotrypsin, calponin, and glial fibrillary acidic protein.

Ghost cell odontogenic carcinoma shows many malignant epithelial islands (containing aberrantly keratinized ghost cells with eosinophilic cytoplasmic cell borders and faint nucleus) in a background of fibrous stroma (similar to calcifying cystic odontogenic tumor and/or dentinogenic ghost cell tumor). Readily identifiable mitoses, necrosis, and osseous destruction with permeation into adjacent tissue are observed. The tumor stains positive for p53 protein as well as PCNA.

Ameloblastic fibrosarcoma consists of a benign epithelial component intermingled within a hypercellular malignant mesenchymal stroma and is considered as the malignant counterpart of ameloblastic fibroma. The tumor shows thin, elongated islands of odontogenic epithelium set in a myxoid mesenchymal stroma under the microscope.

Among *benign odontogenic tumors*, ameloblastoma is a slow-growing but aggressive benign neoplasm with a high recurrence rate and a predilection for the posterior mandible. The tumor presents as a unilocular or multilocular radiolucency around the crown of an unerupted tooth. Histologically, ameloblastoma is characterized by islands or strands of odontogenic epithelium with mature connective tissue stroma. The tumor has EGFR overexpression, which correlates with SOX2 expression in ameloblastic carcinoma in contrast to ameloblastoma.

Among *odontogenic allied lesions*, ossifying fibroma is a well-demarcated firm lesion that contains fibrous tissue with a mineralized component (e.g., woven bone, lamellar bone, and acellular to poorly cellular basophilic and smoothly contoured deposits).

Among *odontogenic cysts*, KCOT is a relatively common cystic neoplasm of the jaws (with the mandibular posterior body and ascending ramus accounting for half of the cases), representing 3%–11% of all gnathic cystic lesions. Radiographically, KCOT displays a well-defined, frequently corticated, uni- or multilocular radiolucency around the crown of an impacted tooth. Histologically, KCOT shows a cystic structure lined by parakeratinized stratified squamous epithelium (typically 6-8 cell layers thick, with prominent, hyperchromatic and palisaded basal cells), with satellite (or daughter) cysts commonly found in the adjacent connective tissue wall.

6.7 Treatment

Treatment for odontogenic tumors, allied lesions, and cysts includes surgery, radiotherapy, and chemotherapy [7,8].

Surgery can be performed through enucleation, enucleation and curettage or peripheral ostectomy, marsupialization, partial or total maxillectomy or mandibulectomy, segmental resection, neck dissection, and cryosurgery after lesion removal. Interestingly, enucleation alone or mandibulectomy is largely recurrence-free, and enucleation and tooth extraction has a recurrence rate of 43.5%. Surgery under general anesthesia gives a recurrence rate of 33%, and that under local anesthesia has no recurrence.

Radiotherapy for odontogenic tumors and lesions typically consists of a mean dose of 64 Gy delivered by intensity modulated radiation therapy (IMRT).

Chemotherapy may comprise cisplatin plus paclitaxel. Inhibitors of mutated *BRAF* (vemurafenib and dabrafenib) may be considered for treating ameloblastoma with *BRAF* mutation, while the MEK inhibitor trametınıb holds promise against mutant NRAS-driven tumors.

6.8 Prognosis

In general, patients with odontogenic tumors undergoing surgery alone have a better survival rate than those receiving adjuvant radiotherapy and/or chemotherapy. Patients with maxillary ameloblastoma tumors carry poor prognosis. The estimated 2-year and 5-year overall survival rates for PIOSCC are 68.9% and 38.8%, respectively.

Although a benign tumor, ameloblastoma behaves unpredictably, and the maxillary tumors carry the worst prognosis. Therefore, patients with ameloblastoma should have a life-time follow-up. Metastasizing ameloblastoma showing indolent but persistent growth (similar to ameloblastoma) often occurs after treatment of the primary jaw tumor, with the lungs (78%) being the most common site of metastasis.

References

1. Barnes L, Eveson JW, Reichart P, Sidransky D. *World Health Organization Classification of Tumours: Pathology and Genetics, Head and Neck Tumours.* Lyon: IARC Press; 2005.
2. Philipsen HP, Reichart PA. Classification of odontogenic tumor. A historical review. *J Oral Pathol Med.* 2006;35:525–9.
3. Richardson MS, Muller S. Malignant odontogenic tumors: An update on selected tumors. *Head Neck Pathol.* 2014;8(4):411–20.
4. Imran A, Jayanthi P, Tanveer S, Gobu SC. Classification of odontogenic cysts and tumors—Antecedents. *J Oral Maxillofac Pathol.* 2016;20(2): 269–71.
5. Bilodeau EA, Prasad JL, Alawi F, Seethala RR. Molecular and genetic aspects of odontogenic lesions. *Head Neck Pathol.* 2014;8(4):400–10.
6. Panda S, Sahoo SR, Srivastav G, Padhiary S, Dhull KS, Aggarwal S. Pathogenesis and nomenclature of odontogenic carcinomas: Revisited. *J Oncol.* 2014;2014:197425.
7. Wright JM, Odell EW, Speight PM, Takata T. Odontogenic tumors, WHO 2005: where do we go from here? *Head Neck Pathol.* 2014;8(4):373–82.
8. Sánchez-Burgos R, González-Martín-Moro J, Pérez-Fernández E, Burgueño-García M. Clinical, radiological and therapeutic features of keratocystic odontogenic tumours: a study over a decade. *J Clin Exp Dent.* 2014;6(3):e259–64.

7
Oral Cavity Cancer

7.1 Definition

Various tumors are known to affect the oral cavity (including the lips), the most important of which are squamous cell carcinoma (SCC), verrucous carcinoma, basal cell carcinoma (BCC), and lymphoma [1].

SCC typically originates from the red lip (especially the lower lip), with a possibility of neck metastases and accounts for >90% of oral cancer. It begins in flat, scalelike squamous cells of the epithelium that covers the mouth and throat. While its earlier form (called carcinoma *in situ*) is present only in the epithelium (the outer layer of cells), its later form (invasive squamous cell carcinoma) grows into deeper layers of the oral cavity or oropharynx.

Verrucous carcinoma is a less common type of SCC that makes up <5% of all oral cancer. As a low-grade cancer, it can grow deeply into surrounding tissue but rarely spreads to other parts of the body. However, without surgical removal, ordinary SCC may develop within some verrucous carcinomas and facilitate the spread to other parts of the body.

BCC often arises from the white lip (especially the upper lip). It is a much less common oral cavity cancer and more amenable to treatment (excision) than SCC.

Lymphoma develops in the tonsils and base of the tongue containing immune system (lymphoid) tissue.

In addition, minor salivary gland tumors may appear on the hard palate, lips and buccal mucosa (see Chapter 9). Moreover, some benign (non-cancerous) tumors and tumor-like conditions (e.g., fosinophilic granuloma, fibroma, granular cell tumor, keratoacanthoma, leiomyoma, osteochondroma, lipoma, schwannoma, neurofibroma, papilloma, condyloma acuminatum, verruciform xanthoma, pyogenic granuloma, rhabdomyoma, and odontogenic tumors) may also affect the mouth or throat. These non-cancerous tumors are generally not life-threatening, and can be removed completely by surgery without recurrence.

7.2 Biology

The oral cavity is an oval-shaped structure demarcated by the lips anteriorly, the cheeks laterally, the floor of the mouth inferiorly, the palate superiorly, and the oropharynx posteriorly. It consists of the lips, the inside lining of the lips, the cheeks (buccal mucosa), the lining of the inside of the cheeks, the teeth, the front two-thirds of the tongue, the upper and lower gums (gingiva), the bottom of the mouth under the tongue, the bony roof of the mouth (hard palate), and the small area behind the wisdom teeth (retromolar trigone).

The primary functions of the oral cavity are to act as the entrance to the alimentary tract, to initiate the digestive process by salivation and propulsion of the alimentary bolus into the pharynx, and to serve as a secondary respiratory conduit, a site of sound modification for the production of speech, and a chemosensory organ.

Involved in the physical breaking down of food, the oral cavity is lined by a protective, nonkeratinized, stratified squamous epithelium, which also covers the inner surface of the lips. Other parts of the oral cavity are lined by masticatory mucosa (gingiva and hard palate), mucosa (the cheeks, alveolar mucosal surface, floor of the mouth, inferior surface of the tongue, soft palate), and a specialized mucosa (dorsal surface of the tongue). However, the outer lip (outer vermilion) is lined with skin, with its epidermis comprising stratified squamous, keratinized epithelium,

The nonkeratinized epithelium is thicker than keratinized epithelium and is composed of the stratum basale (a single layer of cells on basal lamina), the stratum spinosum (several cells thick), and the stratum superficiale (the most superficial layer). The masticatory mucosa possesses a keratinized and, in some areas, a parakeratinized stratified squamous epithelium, whereas the mucosal epithelium contains keratinocytes, Langerhans cells, melanocytes, and Merkel cells.

Oral cavity cancer (e.g., SCC and verrucous carcinoma) usually begins in the squamous cells with the formation of various premalignant (or precancerous) lesions such as leukoplakia (a white/gray patch or plaque that does not rub off), erythroplakia (a flat or slightly raised, bright red velvety patch that often bleeds easily if scraped), erythroleukoplakia (or speckled leukoplakia, a mixed red and white patch), oral lichen planus (particularly the erosive type), oral submucous fibrosis (which is characterized by limited opening of mouth and a burning sensation on eating spicy food and occurs almost exclusively in India and Indian communities living abroad), etc. Based on the degree of dysplasia, premalignant lesions (e.g., leukoplakia, erythroplakia, oral lichen planus, and oral sub-mucous fibrosis) are classified by the WHO into mild, moderate, severe, and carcinoma *in situ*.

7.3 Epidemiology

As the third most common cancer after stomach and cervical cancer and the largest group of head and neck cancer, oral cancer mainly occurs in people of >50 years of age (90%) and rarely in people <50 years of age (10%). Men tend to be more often affected than women.

Oral cancer is particularly prevalent in Sri Lanka, India, Pakistan, Bangladesh, and Brazil as well as some areas of northern France and Hungary.

7.4 Pathogenesis

Risk factors for oral cavity cancer include (i) tobacco smoking/chewing, (ii) alcohol drinking, (iii) betel quid or gutka chewing, (iv) diet low in beta-carotene–rich vegetables and citric fruits, (v) poor oral health, (vi) infection with *Candida albicans*, human herpes virus, and human papillomavirus, (vii) exposure to sunlight or UV, (viii) premalignant lesions and other oral conditions, and (ix) immunosuppression. Genetic changes observed in oral cavity cancer include mutations in chromosomes 3, 9 (*P16*), 11 (*PRAD1*), and 17 (H-*ras*) [2].

7.5 Clinical features

Clinical signs of oral cavity cancer range from (i) a sore (or a granular ulcer with fissuring or raised exophytic margins) on the lips or in the mouth that persists for 3 weeks or longer; (ii) a lump or thickening with abnormal supplying blood vessels on the lips, gums, mouth, or neck; (iii) a white or red patch (erythroplakia, leukoplakia, speckled leukoplakia, or verrucous leukoplakia) on the gums, tongue, or lining of the mouth; (iv) bleeding, pain, or numbness in the lips, mouth, chin, or cheek; (v) change in voice; (vi) loose teeth or dentures; (vii) trouble chewing or swallowing or moving the tongue or jaw; (viii) swelling of jaw; (ix) sore throat or a feeling that something is caught in the throat; to (x) weight loss.

Specifically, oral SCC often looks like scaly red patches, open sores, elevated growths with a central depression, or warts; which may crust or bleed. It can become disfiguring and sometimes deadly if allowed to grow. In late-stage oral SCC, symptoms may include an indurated area, paresthesia or dysesthesia of the tongue or lips, airway obstruction, chronic serous otitis media, otalgia, trismus, dysphagia, cervical lymphadenopathy, persistent pain or referred pain, and altered vision. Oral BCC often looks like open sores, red patches, pink growths, shiny bumps, or scars. BCC almost never spreads (metastasizes) beyond the original tumor site.

7.6 Diagnosis

Oral SCC often looks like scaly red patches, open sores, elevated growths with a central depression, or warts on the lip or lateral part of the tongue. Therefore, application of the mnemonic RULE (red, ulcerated, lump, extending for 3 or more weeks) is valuable for its clinical diagnosis. Oral SCC may crust, bleed, and become disfiguring and sometimes deadly if allowed to grow. Microscopically, the tumor may show verrucoid growth pattern, with atypia at the base, and irregular and infiltrative stromal invasion. It stains positive for epithelial membrane antigen (EMA) and variably positive for BCL-2 but negative for Ber-EP4 and smooth muscle actin (SMA) [3].

Verrucous carcinoma is a rare, low-grade variant/subtype of SCC, which manifests as an ulcerating, fungating, or polypoid mass (1–10 cm) with sinus tracts opening onto skin, and occasional invasion into adjacent soft tissue and bone. Microscopically, the tumor contains well-differentiated hyperplastic squamous epithelium with orderly maturation (upwards and downwards), hyperplastic surface papillae with keratin, also in invaginations; broad, blunt, downward-pushing rete pegs; minimal atypia; presence of mitotic activity; and lymphoplasmacytic infiltration in the lamina propria [4].

Oral BCC presents as open sores, red patches, pink growths, shiny bumps, or scars, with ulceration and crusting. Microscopically, the tumor shows nests of cells with clefting from overlying surface epithelium, pallisading nuclei, and minimal pleomorphism. It stains positive for Ber-EP4 and BCL-2, variable for SMA, but negative for EMA.

Lymphoma is a soft, bulky mass covered by normal or ulcerated mucosa. Microscopically, it shows monomorphic population of immunoblasts with no minimal plasmacytic differentiation; starry sky appearance at low power due to tingible body macrophages; large tumor cells with abundant, basophilic cytoplasm and occasional paranuclear hofs; eccentric, round/oval nuclei with one or more prominent nucleoli; presence of mitotic figures and apoptosis; and infiltration of tumor cell in a large, cohesive mass with a relatively well-delineated advancing edge. The tumor stains positive for EBV, CD38, CD138, MUM1 (100%), intracytoplasmic IgG (50%), and variable light chain restriction but negative for HIV1 (but adjacent benign T cells were HIV1+), HHV8, CD20, and CD45 [5].

Differential diagnoses for oral cavity cancer include actinic keratosis, dermatologic manifestations of oral leukoplakia (a white, largely non-malignant or non-premalignant plaque resulting from increased keratinization or candidosis), erythroplasia (a red, velvety, non-plaque forming lesion, which is level with or depressed below the surrounding mucosa, and which tends to display frank malignancy or severe dysplasia), lichen planus and mucosal candidiasis.

Using the widely applied TNM system of AJCC, the stages of primary oral cavity tumors are determined on the basis of the tumor size and invasion of deep structures as 0 (carcinoma in situ), I, II, III, IVA, IVB and IVC.

7.7 Treatment

Standard treatments for oral cavity cancer are surgery and radiotherapy. Other new types of treatment (i.e., chemotherapy, hyperfractionated radiation therapy, and hyperthermia therapy) may also be considered [6].

For oral SCC, treatment measures may include surgical removal of the entire tumor followed by radiotherapy and/or chemotherapy (in most cases) or a combination of chemotherapy, radiotherapy, and invasive procedures (in rare, metastatic cases). Reconstructive surgery may be necessary after cancer therapy.

For oral BCC, surgery is undertaken through curettage and electrodesiccation (for small lesions), Mohs micrographic surgery (for removal of a thin layer of tissue containing the tumor), excisional surgery (for removal of the entire growth together with a surrounding border of apparently normal skin), radiation (for direct destruction of tumor tissue by X-ray), cryosurgery (for destruction of tumor tissue by liquid nitrogen freezing), photodynamic therapy (for treatment of superficial or nodular BCC, by a strong blue light in the presence of a light-sensitizing agent), and laser surgery (for destruction of lesions by lasers). Topical medications of imiquimod (for superficial BCC) and 5-fluorouracil (also for superficial BCC) may be administered. Further, oral medicines (e.g., vismodegib [Erivedge™, for extraordinarily rare cases of metastatic BCC or locally advanced BCC] and sonidegib [Odomzo® for patients with locally advanced BCC]) may be considered [7,8].

7.8 Prognosis

Patients with oral cavity cancer that is small and still confined to the primary site (Stage I, <2 cm) generally have a better prognosis than those with cancers at a late stage (Stages III and IV, >4 cm or with spread). When all stages of initial diagnosis are considered, the 5-year survival is 55%–63%.

References

1. Montero PH, Patel SG. Cancer of the oral cavity. *Surg Oncol Clin N Am.* 2015; 24(3): 491–508.
2. Gasche JA, Goel A. Epigenetic mechanisms in oral carcinogenesis. *Future Oncol.* 2012; 8(11): 1407–25.
3. Wolff KD, Follmann M, Nast A. The diagnosis and treatment of oral cavity cancer. *Dtsch Arztebl Int.* 2012; 109(48): 829–35.
4. Pathologyoutlines.com. *Verrucous carcinoma.* http://www.pathologyoutlines.com/topic/skintumornonmelanocyticverrucousscc.html. Accessed December 15, 2016.
5. Pathologyoutlines.com. *Lymphoma.* http://www.pathologyoutlines.com/topic/oralcavitylymphoma.html. Accessed December 15, 2016.
6. Chinn SB, Myers JN. Oral cavity carcinoma: Current management, controversies, and future directions. *J Clin Oncol.* 2015; 33(29): 3269–76.
7. De Felice F, Musio D, Terenzi V, et al. Treatment improvement and better patient care: Which is the most important one in oral cavity cancer? *Radiat Oncol.* 2014; 9: 263.
8. PDQ Adult Treatment Editorial Board. *Lip and Oral Cavity Cancer Treatment (PDQ®): Health Professional Version.* PDQ Cancer Information Summaries. Bethesda, MD: National Cancer Institute; 2002–2015.

8
Pharyngeal Cancer

8.1 Definition

Pharyngeal cancer can be separated into nasopharyngeal, oropharyngeal, and hypopharyngeal categories according to the site of development [1].

Nasopharyngeal cancer (or *nasopharyngeal carcinoma*, NPC) is a malignant neoplasm, arising from the squamous epithelium of the nasopharynx. NPC consists of two types: keratinizing and nonkeratinizing squamous cell carcinomas (SCC). The keratinizing SCC is also known as well-differentiated NPC (WHO Type I). The nonkeratinizing SCC is further divided into differentiated NPC (or moderately differentiated NPC, WHO Type II) and undifferentiated NPC (or poorly differentiated NPC, WHO Type III). Interestingly, the nonkeratinizing undifferentiated NPC was previously erroneously referred to as *lymphoepithelioma*, as the tumor is of epithelial origin, and the lymphocytes present are not neoplastic.

Oropharygeal cancer (or *oropharygeal SCC*) usually begins in the squamous cells of the oropharynx and consists of two categories: HPV positive (human papillomavirus–infected), and HPV negative (alcohol or tobacco-related).

Hypopharyngeal cancer forms in the squamous cells of the hypopharynx (laryngopharynx) and comprises squamous cell (epidermoid) carcinoma, basaloid squamoid carcinoma, spindle cell (sarcomatoid) carcinoma, small-cell carcinoma, nasopharyngeal-type undifferentiated carcinoma (lymphoepithelioma), and carcinoma of the minor salivary glands (see Chapter 9).

8.2 Biology

The pharynx (plural: *pharynges*; commonly referred to as the *throat*, which also includes the larynx) is a hollow tube (funnel) of about 12.5 cm long that starts behind the nose and ends at the top of the esophagus and larynx (voice box). The pharynx may be divided into three sections: the nasopharynx (the upper portion of the pharynx, which sits behind the nose where the nasal passages and auditory tubes join the remainder of the upper respiratory tract), the oropharynx (the middle portion of the pharynx, which lies behind the oral cavity, opens anteriorly into the mouth, and includes the back one-third of the tongue, soft palate, side and back walls of the throat, and the tonsils), and the laryngopharynx (the caudal part of the pharynx, also known as *hypopharynx*, which diverges into the respiratory [larynx] and digestive [esophagus] pathways, and includes the pyriform sinuses, posterior pharyngeal wall, and post-cricoid area).

Anatomically, the pharynx is made up of two sets of pharyngeal muscles (arranged as an inner longitudinal layer and an outer circular layer) that are innervated by the pharyngeal plexus. The surface of the nasophar-ynx is lined by two main types of epithelia, that is, a stratified squa-mous epithelium (about 60% of nasopharyngeal epithelium) and a pseudostratified columnar respiratory-type epithelium (about 40% of nasopharyngeal epithelium, consisting of ciliated cells, goblet cells, and basal cells). The surface of the oropharynx and hypopharynx is covered by a nonkeratinizing stratified squamous epithelium (consisting of the basal layer, stratum spinosum, and squamous cells). Despite their com-mon origin (ectoderm), the nonkeratinizing stratified squamous epithe-lium is present in the wet surfaces of oral cavity, esophagus, and vagina, whereas the keratinizing stratified squamous epithelium occurs in the dry surface (skin).

Pharyngeal cancer mostly evolves from the squamous cells that line the moist surface of the pharynx. Keratinizing squamous cell carcinoma (or well-differentiated NPC), nonkeratinizing differentiated NPC (or moderately differentiated NPC), and nonkeratinizing undifferentiated NPC (or poorly differentiated NPC) often arise on the lateral wall of the nasopharynx (fossa of Rosenmüller) and occasionally the superior posterior wall, accounting for ~25%, ~12%, and >60% of all NPC, respectively.

8.3 Epidemiology

NPC has annual incidences of 5–30 per 100,000 and mainly affects middle-aged persons (especially in the fourth to sixth decades, males) in Asia and children in Africa. About 95% of NPC cases involve moderately and poorly differentiated pathological types. Oropharyngeal cancer affects people of >45 years of age, although an increasing number of people aged 20–30 have been diagnosed in recent years. Hypopharyngeal cancer typically occurs in males aged 55–70 years with a history of tobacco and/or alcohol use and rarely in people of <30 years. The tumor is present in the pyriform sinuses (65%–85%), posterior pharyngeal wall (10%–20%), and postcricoid area (5%–15%). Women tend to develop more postcricoid cancer (due to Plummer–Vinson syndrome) than men.

8.4 Pathogenesis

Risk factors for NPC include Asian ancestry, consumption of salted vegetables, fish, and meat, Epstein–Barr virus infection, heavy alcohol use, cigarette smoking, and occupational exposure to chemical fumes, smoke, and formaldehyde. Risks for oropharyngeal cancer are heavy tobacco use (10 packs a day), excessive alcohol intake, infection with human papillomavirus (especially Types 16 and 18), diet low in fruits and vegetables, drinking maté, chewing betel quid, and occupational exposure (e.g., textile fibers, woodworking, and nickel refining, etc.). Finally, risks for hypopharyngeal cancer are tobacco smoking or chewing, heavy alcohol use, diet containing inadequate nutrients, and Plummer–Vinson syndrome.

At the molecular level, NPC is linked to gains in chromosome arms 12p, 1q, 11q, 12q, and 17q and losses in 3p, 9p, 11q, 13q, and 14q. Indeed, chromosomal loss of 3p and gain of chromosome 12 have been recognized as important hallmarks for an early event in NPC. Oropharyngeal cancer is associated with alterations in *TP53* (17p13.1), *NOTCH1* (9p34.3), *CDKN2A* (9p21.3), *FAT1* (4q35.2), *PTEN* (10q23.3), *FBXW7* (4q31.3), *HRAS* (11p15.5), *PIK3CA* (3q26.32), and *EGFR* (7p12). Hypopharyngeal cancer often harbors chromosomal gain on 11q13 and genetic changes relating to EGFR, ERCC1, E-cadherin, β-catenin, epiregulin and amphiregulin, and TP53 [2].

8.5 Clinical features

Patients with pharyngeal cancer may display troubled breathing (chronic cough, wheezing), speaking (hoarse or raspy sound, nasal twang), swallowing, or hearing; lump in the neck, sore throat, or ear pain; headache; trismus; otitis media; cranial nerve palsy (paralysis); nasal obstruction or bleeding; bone pain; organ dysfunction; and paraneoplastic syndrome of osteoarthropathy (diseases of joints and bones).

In advanced NPC, a cluster of symptoms (collectively referred to as *Trotter's syndrome*) may appear, ranging from unilateral conductive deafness (due to middle ear effusion), trigeminal neuralgia (due to perineural spread), nasal regurgitation (due to soft palate immobility), to difficulty opening mouth. Oropharyngeal cancer often presents with a lump in the neck and sore throat; and hypopharyngeal cancer induces sore throat and ear pain.

8.6 Diagnosis

Keratinizing squamous cell carcinoma (or well-differentiated NPC) is a variant of NPC showing keratinization. Macroscopically, the tumor varies from mucosal bulge to an infiltrative mass lesion, with cervical lymph node metastasis. Histologically, the tumor displays irregular islands separated by desmoplastic stroma and obvious squamous differentiation with intercellular bridges and areas of keratinization along with hyperchromatic nuclei. Immunohistochemically, the tumor is positive for pan-cytokeratin and high molecular weight cytokeratin (CK), focally positive for epithelial membrane antigen (EMA), and variably positive for EBV [3].

Nonkeratinizing differentiated NPC (or *moderately differentiated NPC*) may appear as a smooth mucosal bulge, raised nodule with or without surface ulceration, infiltrative mass lesion or an occult lesion. Histologically, the tumor shows interconnecting cords or trabeculae, well-defined cell borders and variable intercellular bridges, and lymphoplasmacytic infiltrate in background stroma. The tumor stains positive for pan-cytokeratin and high molecular weight cytokeratin (CK), variably positive for low molecular weight CK, but negative for CK 7 and CK 20 [3].

Nonkeratinizing undifferentiated NPC (or *poorly differentiated NPC*) appears similar to nonkeratinizing differentiated NPC and displays either syncytial arrangement of cohesive cells with indistinct cell margins (Regaud pattern) or diffuse cellular infiltrate of noncohesive cells (Schminke pattern). The tumor cells show moderate eosinophilic to amphophilic cytoplasm, round nuclei, prominent eosinophilic nucleoli, and vesicular chromatin; insignificant keratinization; presence of apoptosis and brisk mitotic activity as well as occasional non-neoplastic lymphoplasmacytic infiltrate; and absence of necrosis. The tumor is strongly positive for pan-cytokeratin and high molecular weight CK, weakly positive for low molecular weight CK, and positive for EBV (75-100%), but negative for CK7, CK20, and p16 [3].

Basaloid squamous cell carcinoma is a variant of SCC with distinct basaloid morphology, and appears as a deeply invasive, large bulky ulcerated fungating mass. Histologically, the tumor shows solid, cribriform, or microcystic nests, strands, trabeculae, or lobules of tumor cells, with strands of tumor cells connected to overlying squamous epithelium; round to oval, hyperchromatic nuclei with peripheral palisading and central comedo-type necrosis; many mitotic figures; microcystic pattern with basophilic material; areas of stromal hyalinization; and absence of true lumens. The tumor stains positive for AE1/AE3, CK14, CK19, EMA, p53, bcl2, and laminin, but negative for neuroendocrine markers, smooth muscle actin (SMA) and S100 (rarely weakly positive).

Spindle cell (sarcomatoid) carcinoma is a biphasic tumor composed of an SCC and a malignant spindle cell component with a mesenchymal appearance, but of epithelial origin. Macroscopically, the tumor has a polypoid appearance with a mean size of 2 cm, and frequent surface ulceration. Histologically, the tumor shows spindle cells (of plump fusiform, or rounded and epithelioid shape) arranged in storiform, interlacing bundles or fascicles, and herringbone; hypocellular areas with dense collagen deposition; mild to moderate pleomorphism, without a severe degree of anaplasia; rare occurrence of metaplastic or frankly neoplastic cartilage or bone; minor to inconspicuous SCC component with the sarcomatoid part dominating. The individual neoplastic spindle cells react variably with keratin (AE1/AE3), EMA, and CK18.

Small-cell carcinoma is a high-grade neuroendocrine tumor that often arises in the nasal cavity and extends into the maxillary or ethmoid sinuses. Histologically, the tumor shows sheets or ribbons of oval to spindle-shaped cells with high nuclear–cytoplasmic ratio, dense chromatin, and inconspicuous nucleoli; nuclear molding, mitoses, necrosis, angiolymphatic, and perineural invasion; and rosettes. The tumor stains positive for chromogranin, synaptophysin, neuron specific enolase, cytokeratin AE1/AE3 (paranuclear, punctuate pattern), and EMA, but negative for TTF1, S100, and EBV.

8.7 Treatment

Current treatments for the early stages of pharyngeal cancer include radiotherapy and surgery, with surgery reserved for patients who fail to respond to radiotherapy [4–6]. Radiotherapy may be undertaken externally (e.g., fast neutron radiation therapy and photon-beam radiation therapy) or internally (e.g., internal radiation therapy).

In more advanced cases without evidence of spreading, chemotherapy may be given in combination with surgery and/or radiation. However, if pharyngeal cancer has already spread to other parts of the body, chemotherapy may be given alone.

Chemotherapy may be systemic or regional. In systemic chemotherapy, drugs are administered orally or injected into a vein or muscle. In regional chemotherapy, drugs are placed directly into the cerebrospinal fluid, an organ, or a body cavity (e.g., the abdomen).

8.8 Prognosis

The 5-year survival rate for patients with NPC is about 90% (Stage I), 70% (Stage II), 55% (Stage III), and 30% (Stage IV). The 5-year survival rate for patients with oropharyngeal cancer is 83% (early stages), 62% (with spreading to surrounding tissues or organs and/or the regional lymph nodes), and 38% (with spreading to a distant part of the body). The 5-year survival rate for patients with hypopharyngeal cancer is 53% (early, localized stage), 36%–39% (Stages II–III, with spreading to nearby areas and/or lymph nodes), and 24% (with spreading to distant parts of the body).

References

1. Barnes L, Eveson JW, Reichart P, Sidransky D. *Pathology and Genetics of Head and Neck Tumours. World Health Organization Classification of Tumours.* Lyon: IARC Press; 2005.
2. Lo KW, Chung GT, To KF. Deciphering the molecular genetic basis of NPC through molecular, cytogenetic, and epigenetic approaches. *Semin Cancer Biol.* 2012; 22(2): 79–86.
3. Pathologyoutlines.com. Nasopharyngeal. carcinoma. http://www.pathologyoutlines.com/topic/nasalnasopharyngealgeneral.html; accessed December 18, 2016.
4. PDQ Adult Treatment Editorial Board. *Hypopharyngeal Cancer Treatment (PDQ®): Patient Version.* PDQ Cancer Information Summaries. Bethesda, MD: National Cancer Institute; 2002–2016.
5. PDQ Adult Treatment Editorial Board. *Nasopharyngeal Cancer Treatment (PDQ®): Patient Version.* PDQ Cancer Information Summaries. Bethesda, MD: National Cancer Institute; 2002–2016.
6. PDQ Adult Treatment Editorial Board. *Oropharyngeal Cancer Treatment (PDQ®): Patient Version.* PDQ Cancer Information Summaries. Bethesda, MD: National Cancer Institute; 2002–2016.

9
Salivary Gland Tumors

9.1 Definition

Tumors of the salivary glands comprise a heterogeneous group of neoplasms with more than 24 different histologic subtypes. Of these, the most commonly occurring malignant salivary gland tumors are mucoepidermoid carcinoma (MEC, 13% of all salivary gland tumors and 35% of malignant salivary gland tumors), adenoid cystic carcinoma (ACC, 8% of all salivary gland tumors), adenocarcinoma NOS, salivary duct carcinoma, acinic cell carcinoma, and polymorphous low-grade adenocarcinoma (PLGA). The most common benign salivary gland tumors are pleomorphic adenoma (68% of all salivary gland tumors and 65% of parotid gland tumors) and Warthin tumor (or *papillary cystadenoma lymphomatosum*, 10% of all salivary gland tumors) [1].

9.2 Biology

The salivary glands are relatively small organs involved in the making of saliva, which is composed of water, electrolytes (sodium, potassium, chloride, and bicarbonate), digestive enzymes (salivary amylase and lingual lipase), proteins (mucin, IgA, lysozyme, and defensins), and metabolic wastes (uric acid, urea). The saliva helps cleanse the mouth, moisten and compact food into a round mass called a *bolus*, break down starch, and protect the mouth and throat against microbial infections.

Based on their location, the salivary glands are divided into (i) the major (or extrinsic) salivary glands (situated outside the oral cavity) and (ii) the minor (or intrinsic) salivary glands (located within the oral cavity).

The major salivary glands comprise three main pairs: (i) the parotid glands (the large, triangle-shaped glands located anterior to the ear between the skin and masseter muscle, with the main duct opening into the vestibule next to the second upper molar); (ii) the submandibular glands (the walnut-sized glands located below the jawbone, with the duct opening underneath the tongue at the base of the lingual frenulum); and (iii) the sublingual glands (the small, almond-shaped glands located in front of the submandibular gland under the tongue, with 10–20 ducts opening into the floor of the mouth).

The minor salivary glands consist of 500–1000 small glands (of 1–5 mm in size) that are separated from one another by connective tissue and that are scattered within the mucosa and submucosa of the lips, floor of the mouth, uvula, hard and soft palates, retromolar trigone, posterior tongue, pharynx, larynx, and paranasal sinuses. However, the anterior hard palate and gingivae are devoid of minor salivary glands.

Structurally, a salivary gland is made up of four segments: acinus, intercalated duct, striated duct, and excretory duct, with two types of cells present in each segment: abluminal cells (as myoepithelial cells in the acinus and intercalated duct, and basal cells in the striated and excretory ducts) and luminal cells (as acinar cells in the acinus, and ductal cells in the striated duct and excretory duct). The acinar cells in the acinus may be either serous or mucous, with the serous cells producing a watery secretion containing ions, enzymes, and mucin and the mucous cells producing mucus. The parotid glands are mostly serous, the submandibular glands are mixed (predominantly serous), the sublingual glands are mixed (predominantly mucous), and the minor salivary glands are mixed (predominantly mucous).

Possibly arising from progenitor cells residing within the ductal tributaries of the salivary glands, salivary gland tumors can be divided into luminal (acinar and ductal cells) and abluminal (myoepithelial and basal cells) types [2]. Myoepithelial cells are stellate shaped with cytoplasmic processes embracing the acini, or spindle shaped surrounding the intercalated ducts. As a dual epithelial and smooth muscle phenotype, myoepithelial cells produce an extracellular matrix such as basement membrane materials and myxoid substances leading to diverse histology of salivary gland tumors. Basal cells of the striated ducts, excretory ducts, and salivary ducts differ ultrastructurally from myoepithelial cells in the absence of myofilaments. Salivary gland tumors containing predominantly myoepithelial cells are considered biologically low grade, while those devoid of myoepithelium are considered high grade.

Of all salivary gland tumors, about 80%–85% originate from the parotid glands, 10% from the submandibular glands, <1% from the sublingual glands, and 5% from the minor salivary glands (mostly involving the palate). Further, about 25% of parotid tumors, 40% of submandibular tumors, >90% of sublingual gland tumors, and 50% of palate tumors are malignant. Common tumors of the minor salivary glands include ACC, PLGA and MEC, which often appear on the hard palate (60%), lips (25%), and buccal mucosa (15%) [3].

Apart from salivary gland tumors, pulmonary salivary gland-like tumors may arise from the uncontrolled cell division (mitosis) of mutated cancer stem cells in the epithelium of the submucosal bronchial glands (which are the pulmonary equivalent of minor salivary glands). These tumors are morphologically identical to salivary gland tumors, with MEC involving the lung parenchyma, and ACC affecting the lateral and posterolateral wall near the junction of the cartilaginous and membranous portions of the proximal one-third or distal one-third of the trachea.

9.3 Epidemiology

Salivary gland tumors are relatively rare, with annual incidence of 3 per 100,000. They account for <5% of head and neck cancers and about 0.5% of all malignancies. The mean ages of patients with malignant salivary gland tumors range from 48 to 58 years. There is a slight male predominance (with a male-to-female ratio of 1.2:1).

A recent survey suggested that ACC (42.1%), MEC (27.6%), adenocarcinoma not otherwise specified (13.6%), polymorphous low-grade adenocarcinoma (3.2%), papillary cystic adenocarcinoma (3.2%), acinic cell carcinoma (2.7%), carcinoma ex pleomorphic adenoma (2.3%), basal cell adenocarcinoma (1.4%), squamous cell carcinoma (1.4%), mucinous adenocarcinoma (0.9%), salivary duct carcinoma (0.9%), oncocytic carcinoma (0.5%), and sebaceous carcinoma (0.5%) represent the main historic types of malignant salivary gland tumors in Africa [3].

9.4 Pathogenesis

Risk factors for salivary gland tumors include exposure to ionizing radiation (e.g., previous radiotherapy for head and neck neoplasms), rubber product manufacturing, asbestos mining, plumbing, and woodworking. Genetic alterations have been also observed in MEC (chromosome translocation t[11;19] [q14–21;p12–13], MECT1-MAML2); ACC and hyalinizing clear cell carcinoma (MYB-NFIB and EWSR1-ATF1 gene fusions, respectively); acinic cell carcinoma, cystadenocarcinoma, and adenocarcinoma NOS (ETV6-NTRK3); and salivary duct carcinoma (6q and 17p/17q mutations, chromosome 7 polysomy, 12q amplification, 9p LOH, methylation of MGMT, DAPK, and RASSF1) [2,4].

9.5 Clinical features

Patients with salivary gland tumors often present with a lump (usually painless) in the area of the ear, cheek, jaw, lip, or inside the mouth; trouble swallowing; drainage from the ipsilateral ear, dysphagia, and trismus; facial paralysis (facial numbness or weakness); and persistent facial pain.

9.6 Diagnosis

MEC is the most common malignant tumor of the major salivary glands (with 84%–93% of cases involving the parotid gland). Macroscopically, MEC is a circumscribed mass with gray-white, mucin-filled cysts. Histologically, MEC shows cords, sheets, and clusters of mucous, squamous, intermediate, and clear cells; occasional focal sebaceous cells, goblet-type cells, oncocytic change, inflammatory reaction to extravasated mucin or keratin; and absence of squamous cell carcinoma *in situ*. In low-grade MEC, mucinous and intermediate cells with bland nuclei forming glandular spaces are observed. In high-grade MEC, solid and infiltrative growth pattern of atypical epidermoid and intermediate cells with cytoplasmic clearing and a small number of mucinous cells, along with <20% intracystic component, are noted. Immunohistochemically, low-grade MEC stains positive for CK7 and CK14. Molecularly, MEC is characterized by a t(11;19)(q14–21;p12–13) translocation [5].

ACC (formerly *cylindroma*) is the most common malignant salivary gland tumor of the minor salivary glands. Macroscopically, ACC is a small, poorly circumscribed or encapsulated and infiltrative, 1–8 cm in size, locally aggressive tumor. Histologically ACC demonstrates three growth patterns: tubular, cribriform (classic), and solid (basaloid). The cribriform pattern is the most common and has epithelial cell nests in cylindrical formation. The tubular pattern contains tubular structures lined by stratified cuboidal epithelium. The solid pattern is the least common and of high grade, displaying solid groups of cuboidal cells. Apart from small bland myoepithelial cells with scant cytoplasm and dark compact angular nuclei surrounding pseudoglandular spaces with PAS+ excess basement membrane material and mucin, ACC shows peripheral perineural invasion and small true glandular lumina; absence of squamous differentiation and extensive necrosis (presence of pseudoglandular lumina, true glandular lumina, and perineurial invasion is diagnostic). Dedifferentiated tumor may have irregular tumor islands composed of anaplastic cells with abundant cytoplasm and desmoplastic stroma. ACC stains positive for keratin, CEA, S100, CK7/CK20, etc., but negative for estrogen and progesterone receptors. Molecularly, ACC is associated with deletion and/or translocation on the long arm of chromosome 6 and altered p53 expression [6].

PLGA (preferably called polymorphous adenocarcinoma) is the second most common tumor of the minor salivary glands (after ACC). The tumor is nonencapsulated, and demonstrates uniform plump columnar cells with bland nuclei and diverse growth patterns (tubular, cribriform, papillary, solid, fascicular, microcystic, single file, pseudoadenoid cystic [without true lumens], strand-like, mixed). Other features include perineural invasion around small nerves; infiltrative borders; up to 12 mitotic figures per 10 HPF; and rare tumor necrosis. The tumor stains positive for S100, EMA, keratin, muscle specific actin and CEA (focal).

Pleomorphic adenoma (also called benign mixed tumor) is a painless, slow growing tumor mostly found in the parotid gland (90%) and submandibular gland (10%). Macroscopically, the tumor is well-demarcated, partially encapsulated, gray-white, myxoid, rubbery mass of <6 cm with solid cut surface and subtle extensions into adjacent tissue. Histologically, the tumor contains biphasic population of epithelial and mesenchymal cells; glandular or squamous, spindled or oval epithelial cells with large hyperchromatic nuclei; myoepithelial basal layer or overlying pseudoepitheliomatous hyperplasia; myxoid, hyaline stroma; presence of mucin; absence of mitotic figures and necrosis. The ductal component stains positive for CK19, CK14, EMA, CEA, alpha-1-antitrypsin, alpha-1-antichymotrypsin, GCDFP-15, PSA (50%), PAP (50%); the myoepithelial component stains positive for keratin, actin, myosin, other smooth muscle proteins, S100 (particularly in cartilaginous areas), and GFAP. The tumor is negative for amylase and p53. and may contain rearrangements at 8q12 or 12q14–15.

Depending on the size and status of spread, salivary gland tumors are divided into stages I, II, III and IV. Stage I tumor is <2 cm; stage II tumor is >2 cm but <4 cm; stage III tumor is <4 cm and has spread to a single lymph node (<3 cm) on the same side, or >4 cm and/or has spread to soft tissue; stage IV (including stages IVA, IVB, and IVC) tumor is >3 cm and has spread to one lymph node on either or both sides as well as to the skin, jawbone, ear canal, and/or facial nerve of the body.

9.7 Treatment

Complete surgical resection (with adequate free margins) and/or radiotherapy represent the current mainstay treatment options for salivary gland tumors [6]. For tumors of the parotid gland, parotidectomy and postoperative radiotherapy are useful (especially when margins are close or involved, when tumors are large, or when histologic evidence of lymph node metastases is present). Chemotherapy (with cisplatin and 5-fluorouracil or cisplatin, doxorubicin, and cyclophosphamide) is generally reserved for the palliative treatment of symptomatic locally recurrent and/or metastatic disease that is not amenable to further surgery or radiation. Other targeted drugs (e.g., adriamycin, carboplatin, methotrexate, docetaxel, paclitaxel, bacilli

Calmette–Guerin, platinum, and anthracyclines as well as bortezomib, imatinib, cetuximab, gefitinib, and trastuzumab) may also be considered [7,8].

9.8 Prognosis

Early stage tumors in the parotid gland can be usually treated by surgery alone and have a more favorable prognosis than those in the submandibular gland and the sublingual and minor salivary glands. Large bulky tumors or high-grade tumors often require surgical resection in combination with postoperative radiation therapy; they carry a poorer prognosis. MEC has a 5-year overall survival rate of 62.3% and a 5-year disease-free survival rate of 57.2%. ACC has an excellent prognosis, with a 5-year survival rate of about 90%; the overall survival in the pediatric age group is 96% at 5 years, 95% at 10 years, and 83% at 20 years [7].

References

1. Sreeja C, Shahela T, Aesha S, Satish MK. Taxonomy of salivary gland neoplasm. *J Clin Diagn Res.* 2014;8(3):291–3.
2. Dwivedi N, Agarwal A, Raj V, Chandra S. Histogenesis of salivary gland neoplasms. *Indian J Cancer.* 2013;50(4):361–6.
3. Lawal AO, Adisa AO, Kolude B, Adeyemi BF. Malignant salivary gland tumours of the head and neck region: A single institutions review. *Pan Afr Med J.* 2015;20:121.
4. Stenman G. Fusion oncogenes in salivary gland tumors: molecular and clinical consequences. *Head Neck Pathol.* 2013;7 Suppl 1:S12–9.
5. Pathologyoutlines.com website. *Mucoepidermoid carcinoma.* http://pathologyoutlines.com/topic/salivaryglandsMEC.html; accessed December 20, 2016.
6. Pathologyoutlines.com website. *Adenoid cystic carcinoma.* http://pathologyoutlines.com/topic/salivaryglandsadenoidcystic.html; accessed December 20, 2016.
7. PDQ Adult Treatment Editorial Board. *Salivary gland cancer treatment (PDQ®): Health Professional Version.* PDQ Cancer Information Summaries. Bethesda, MD: National Cancer Institute; 2002–2015.
8. Mifsud MJ, Burton JN, Trotti AM, Padhya TA. Multidisciplinary management of salivary gland cancers. *Cancer Control.* 2016;23(3):242–8.

SECTION II
Cardiovascular and Respiratory Systems

10
Cardiac Tumors

10.1 Definition

Arising from various tissues in the heart or heart valves, primary cardiac tumors consist of three categories: (i) benign tumor and tumor-like lesions, (ii) malignant tumors, and (iii) pericardial tumors [1].

Benign tumors and tumor-like lesions of the heart account for about 80% of all cardiac tumors and include rhabdomyoma, histiocytoid cardiomyopathy, hamartoma of mature cardiac myocytes, adult cellular rhabdomyoma, cardiac myxoma, papillary fibroelastoma, hemangioma, cardiac fibroma, inflammatory myofibroblastic tumor, lipoma, and cystic tumor of the atrioventricular node.

Malignant tumors of the heart account for about 15% of all cardiac tumors and comprise angiosarcoma, epithelioid hemangioendothelioma, malignant pleomorphic fibrous histiocytoma/undifferentiated pleomorphic sarcoma, fibrosarcoma and myxoid fibrosarcoma, rhabdomyosarcoma, leiomyosarcoma, synovial sarcoma, liposarcoma, and cardiac lymphomas.

Pericardial tumors of the heart consist of solitary fibrous tumor, malignant mesothelioma, and germ cell tumors.

Besides primary cardiac tumors that begin and stay in the heart, secondary (metastatic) tumors that start in another part of the body (e.g., lung, breast, bone, liver, and kidney) may move (metastasize) to the heart and pericardial tissues. Secondary tumors of the heart are malignant and represent 3%–5% of all cardiac neoplasms [2].

10.2 Biology

Located in the thoracic cavity medial to the lungs and posterior to the sternum, the heart is a muscular organ about the size of a closed fist. Its main functions are to take in deoxygenated blood through the veins, deliver it to the lungs for oxygenation, and then pump it via the arteries to various parts of the body.

Anatomically, the heart sits within a fluid-filled cavity called the *pericardial cavity*, which is covered by a special membrane known as the *pericardium*.

The pericardium consists of a visceral layer (also known as the *epicardium*, which covers the outside of the heart) and a parietal layer (forming a sac around the outside of the pericardial cavity).

The heart wall is composed of three layers: epicardium, myocardium, and endocardium. As the outermost layer of the heart wall, the epicardium (or the visceral layer of the pericardium) is a thin layer of serous membrane that lubricates and protects the outside of the heart. Underneath the epicardium is the myocardium, which is a thick layer of the heart wall that comprises the cardiac muscle tissue. Further below the myocardium is the endocardium, which is a thin, squamous endothelium layer that lines the inside of the heart.

The heart contains four chambers: right atrium, left atrium, right ventricle, and left ventricle. Being smaller with thinner, less-muscular walls than the ventricles, the atria are connected to the veins that carry blood to the heart and thus act as receiving chambers for blood. The ventricles are the larger, stronger pumping chambers that are connected to the arteries, which carry blood away from the heart and act to send it to various parts of the body. Specifically, the right side of the heart maintains pulmonary circulation to the nearby lungs, whereas the left side of the heart pumps blood all the way to the extremities of the body in the systemic circulatory loop.

The heart utilizes a system of one-way valves (including atrioventricular and semilunar valves) to prevent blood from flowing backwards or regurgitating back into the heart. Situated in the middle of the heart between the atria and ventricles, the atrioventricular valves include the tricuspid valve and the mitral valve and only allow blood to flow from the atria into the ventricles. Located between the ventricles and the arteries that carry blood away from the heart, the semilunar valves include the pulmonary valve (which prevents the backflow of blood from the pulmonary trunk into the right ventricle) and the aortic valve (which prevents the aorta from regurgitating blood back into the left ventricle).

The heart is evolved from the embryonic mesodermal germ layer cells that differentiate into the mesothelium, endothelium, and myocardium. Primary cardiac tumors showing differentiation into muscle cells (heart wall) are rhabdomyoma, adult cellular rhabdomyoma, hamartoma of mature cardiac myocytes, and histiocytoid cardiomyopathy; those of pluripotent mesenchymal origin (lining of the heart chambers) are cardiac myxoma and papillary fibroelastoma; those showing differentiation into myofibroblastic cell (heart valves) are cardiac fibroma and inflammatory myofibroblastic tumor; that of vascular tissue origin is hemangioma; and congenital cystic lesion in the atrioventricular node is known as cystic tumor of the atrioventricular node (a mass of conductive tissue located in the right atrium that receives a signal from the sinoatrial node that sets the pace of the heart) [2].

10.3 Epidemiology

Cardiac tumors are rare, with an annual incidence of 1 per 100,000. This may be attributable to the fact that the heart is less exposed to external irritation, and myocytes never divide.

The most common primary cardiac tumors are benign (e.g., cardiac myxoma, papillary fibroelastoma, rhabdomyoma, fibroma, hemangioma, lipoma, and cystic tumor of the atrioventricular node). Frequently detected primary malignant cardiac tumors are angiosarcoma, rhabdomyosarcoma, malignant fibrous histiocytoma, leiomyosarcoma, and unclassified sarcoma. Notable secondary malignant tumors of the heart are melanoma, lung and breast carcinoma, soft-tissue sarcoma, renal cancer, leukemia, and lymphoma.

Whereas cardiac myxomas, lipomatous hyperplasia, mesothelioma, paraganglioma, and sarcoma tumors are identified mainly in adults, rhabdomyoma (often as part of tuberous sclerosis), fibroma, histiocytoid cardiomyopathy, teratoma, hamartoma, and Purkinje tumors are often found in children. Papillary fibroelastoma, hemangioma, and lipoma occur in all age groups [3].

10.4 Pathogenesis

A number of cardiac tumors have been associated with other health conditions such as Carney complex and tuberous sclerosis complex (TSC).

Carney complex is a familial, autosomal dominant syndrome characterized clinically by cardiac and extracardiac (cutaneous) myxomas, pigmented skin lesions (lentigines, ephelides, blue nevi), endocrine neoplasia, and psammomatous melanotic schwannoma. The most common cause of Carney complex relates to mutations in the *PRKAR1A* gene on chromosome 17q22-24, and about 10% of cardiac myxomas harbor mutations in this gene.

TSC is an autosomal dominant syndrome characterized by hamartoma formation in multiple organ systems and mutations in the tuberous sclerosis genes TSC1 and TSC2. About 50% of cardiac rhabdomyoma cases are linked to TSC.

Primary cardiac angiosarcoma contains KDR (G681R) mutation and focal high-level amplification at chromosome 1q encompassing MDM4.

A missense variant (p.R117C) in the *POT1* gene is also noted in cardiac angiosarcoma from T53-negative Li–Fraumeni-like families.

10.5 Clinical features

Patients with benign cardiac tumors often display clinical symptoms related to extracardiac, intramyocardial, or intracavitary abnormalities. Extracardiac symptoms include dyspnea, chest discomfort, and specific manifestations associated with embolism. Other less specific symptoms are fever, chills, cough, dizziness (fainting, lightheadedness), palpitations (rapid heart rate), hypotension (due to bleeding into the pericardium), lethargy, arthralgias, involuntary weight loss, and petechiae. Intramyocardial symptoms typically consist of abnormal heart rhythms or heart murmurs (arrhythmias). Intracavitary symptoms include valvular stenosis, valvular insufficiency, or heart failure, due to tumor obstruction of valvular function and/or blood flow.

Patients with malignant cardiac tumors tend to have more acute symptoms in onset and rapid progression. For example, cardiac sarcomas are responsible for ventricular inflow tract obstruction and pericardial tamponade. Mesothelioma leads to pericarditis or tamponade. Metastatic cardiac tumors may cause sudden cardiac enlargement, tamponade, heart block, arrhythmias, or sudden heart failure, in addition to fever, malaise, weight loss, night sweats, and loss of appetite.

In addition, some patients with certain cardiac tumors (e.g., fibroelastoma and rhabdomyoma) may be asymptomatic.

10.6 Diagnosis

Imaging procedures (e.g., echocardiography, MRI, CT) are commonly used for diagnosis and differentiation of benign and malignant cardiac tumors.

Among primary benign tumors of the heart, cardiac myxoma accounts for 50% of all heart neoplasms and presents as a solitary, polypoid, pale, lobulated, pedunculated mass of up to 15 cm predominantly in the left atrium with a predilection for women (mean, 50 years). Histologically, cardiac myxoma shows complex structures (eg, cords, nests, rings or poorly formed glands, often surrounding blood vessels) formed by stellate or globular myxoma cells with abundant eosinophilic cytoplasm, indistinct cell borders, oval nucleus with open chromatin and indistinct nuclei; abundant mucopolysaccharide (myxoid) ground substance containing chondroitin sulfate and hyaluronic acid; inflammation, hemorrhage, cellular and mitotic activity near surface; fibrosis (41%), calcification (20%), Gamna-Gandy bodies (17%), ossification (8%), extramedullary hematopoiesis (7%), mucin-forming glands (3%), atypia (3%), and thymic rests (1%). The tumor stains positive for CD31, CD34, calretinin (strong, diffuse, cytoplasmic and nuclear staining), S100, and vimentin, but negative for CD68, cytokeratin (except for glandular elements), and Factor VIII (except for surface cells) [4]. Papillary fibroelastoma (also called *giant Lambl's excrescence*) is the second most common benign tumor, mainly occurring in adults (mean, 60 years) and representing 8%–10% of all heart tumors. It appears as a distinctive cluster of yellow-white hairlike projections of up to 1 cm in diameter covering large portions of the valvular surface (80%, particularly aortic and mitral valves) or endocardium (20%) [5]. Rhabdomyoma accounts for up to 90% of primary heart tumors in infants and children (with tuberous sclerosis) and presents as small, firm, gray-white, well-circumscribed myocardial masses (often multiple) of 3–4 cm (up to 10 cm) that protrude into the ventricles [6].

Among primary malignant tumors of the heart, angiosarcoma is a large, infiltrating, pink to dark brown, lobulated mass located in the right atrium of men (mean, 40 years). Histologically, the tumor shows single or clusters of pleomorphic, poorly differentiated spindle cells; anastomosing vascular channels forming dilated sinusoids lined by abnormal spindle or polygonal cells; prominent mitotic activity, hemorrhage and necrosis. The tumor is positive for CD31 (most specific), CD34, reticulin (highlighting vascular channels), thrombomodulin, and factor VIII (20%); but negative for CD45, cytokeratin, and S100 [7]. Rhabdomyosarcoma is the most common cardiac malignancy in infants and children and presents as a bulky, invasive mass of >10 cm with central necrosis, with mitral valve or atrial wall involvement. Being a sarcoma with striated muscle differentiation, it contains cells with cross striations, and stains positive for PAS and desmin.

10.7 Treatment

Standard treatment for primary benign cardiac tumors is excision, that for primary malignant cardiac tumors is palliation (e.g., radiation therapy, chemotherapy, management of complications), and that for metastatic cardiac tumors may include systemic chemotherapy or palliation, pacemaker implantation, anti-arrhythmic drugs, and possibly surgical excision [8].

About half of rhabdomyoma-affected newborns may regress spontaneously and do not require treatment; fibroelastoma may involve valvular repair or replacement; large fibroma may require heart transplantation.

10.8 Prognosis

Surgery is usually curative for benign primary tumors of the heart, with a 3-year survival rate of 95%. However, multifocal rhabdomyomas or fibromas of the heart have poor prognosis, with a 5-year survival rate of 15%, as surgical excision is usually ineffective. Malignant cardiac tumors often have a poor prognosis.

References

1. Travis WD, Brambilla E, Burke AP, Marx A, Nicholson AG. *WHO Classification of Tumours of the Lung, Pleura, Thymus and Heart.* International Agency for Research on Cancer, Lyon; 2015.
2. Amano J, Nakayama J, Yoshimura Y, Ikeda U. Clinical classification of cardiovascular tumors and tumor-like lesions, and its incidences. *Gen Thorac Cardiovasc Surg.* 2013;61(8):435–47.
3. Yuan SM. Fetal primary cardiac tumors during perinatal period. *Pediatr Neonatol.* 2017;58(3):205–10.
4. Pathologyoutlines.com website. Myxoma. http://www.pathologyoutlines.com/topic/hearttumormyxoma.html; accessed December 21, 2016.
5. Pathologyoutlines.com website. Papillary fibroelastoma. http://www.pathologyoutlines.com/topic/hearttumorfibroelastoma.html; accessed December 21, 2016.
6. Pathologyoutlines.com website. Angiosarcoma. http://www.pathologyoutlines.com/topic/hearttumorangiosarcoma.html; accessed December 21, 2016.
7. Pathologyoutlines.com website. Rhabdomyoma. http://www.pathologyoutlines.com/topic/hearttumorrhabdomyoma.html; accessed December 21, 2016.
8. Yin L, He D, Shen H, Ling X, Li W, Xue Q, Wang Z. Surgical treatment of cardiac tumors: a 5-year experience from a single cardiac center. *J Thorac Dis.* 2016;8(5):911–9.

11
Large Cell Carcinoma and Large Cell Neuroendocrine Carcinoma of the Lung

11.1 Definition

Traditionally, primary tumors of the lung are divided into two categories: small cell lung carcinoma (SCLC) and non-small cell lung carcinoma (NSCLC), which represent about 20% and 80% of clinically identified lung tumors, respectively.

The 2015 WHO classification of tumors of the lung, pleura, thymus, and heart separates primary tumors of the lungs into epithelial, mesenchymal, and lymphohistiocytic tumors as well as tumors of ectopic origin. In turn, epithelial tumors are subdivided into 11 types: (i) adenocarcinoma, (ii) squamous cell carcinoma (SCC), (iii) neuroendocrine (NE) tumors (consisting of SCLC, large cell NE carcinoma [LCNEC], carcinoid tumors (typical and atypical)], (iv) large cell carcinoma (LCC), (v) adenosquamous carcinoma, (vi) sarcomatoid carcinoma, (vii) salivary gland–like tumors, (viii) preinvasive lesions, (ix) other and unclassified carcinomas, (x) papillomas, and (xi) adenomas [1]. Among these, adenocarcinoma (35%), SCC (30%), SCLC (20%), and LCC (9%), and LCNEC (3%) account for about 97% of all lung tumors [1].

Large cell carcinoma of lung (LCC, or *large cell lung cancer*; formerly *large cell anaplastic carcinoma* and *large cell undifferentiated carcinoma*) is an undifferentiated NSCLC that contains large, round cells and lacks the cytologic and architectural features of small cell carcinoma and glandular or squamous differentiation.

LCNEC of the lung is defined as NSCLC with NE morphology and positive NE markers, whereas NSCLC with NE morphology but negative NE markers is regarded as large cell carcinoma with NE morphology (LCNEM) [2].

11.2 Biology

The human respiratory system contains two major components, the airway and the lungs. The airway is separated into (i) the extrathoracic (superior) airway, consisting of the supraglottic, glottic, and infraglottic regions, and (ii) the intrathoracic (inferior) airway, consisting of the trachea (windpipe), mainstem bronchi (tubular branches), and 23 bronchial generations (bronchioles), whose main functions are to supply air (oxygen) to, and remove carbon dioxide from, the alveoli (air sacs) and within the lungs.

The lungs are a pair of spongy, air-filled organs located on either side of the chest (thorax). They consist of the lung parenchyma, which is subdivided into lobes and segments (the gross functional subunits of the lungs). The right lung includes ten segments—three in the right upper lobe (apical, anterior, and medial), two in the right middle lobe (medial and lateral), and five in the right lower lobe (superior, medial, anterior, lateral, and posterior). The left lung includes eight segments—four in the left upper lobe (apicoposterior, anterior, superior lingula, and inferior lingula) and four in the left lower lobe (superior, anteromedial, lateral, and posterior).

The outer surface of the lungs is covered by a thin tissue layer called the *visceral pleura*, which is contiguous with (and thinner than) the parietal pleura (lining the inside of the chest cavity) at the hilum of each lung. The visceral pleura forms invaginations (called *fissures*) into both lungs, with two complete fissures in the right lung and one complete fissure and one incomplete fissure in the left lung. The fissures separate the different lung lobes.

Structurally, the trachea is a cartilaginous/fibromuscular tube of 10–12 cm in length (including 2–4 cm extrathoracic and 6–9 cm intrathoracic) and 11–25 mm in diameter. The tracheal wall is composed of four layers: mucosa, submucosa, cartilage/muscle, and adventitia. The mucosa contains a ciliated pseudostratified columnar epithelium and numerous mucus-secreting goblet cells (which rest on a basement membrane with a thin, collagenous lamina propria). The submucosa contains seromucous glands. The adventitia contains C-shaped, hyaline cartilage rings interconnected by connective tissue. In the posterior tracheal wall, cartilage is replaced by a thin band of smooth muscle (the trachealis muscle).

Similar to the trachea, the bronchi (whose wall thickness is about one-sixth to one-tenth of the airway diameter) are composed of cartilaginous and fibromuscular elements. The epithelium of the bronchus transitions from pseudostratified columnar ciliated epithelium (with goblet cells) into a simple columnar ciliated epithelium first and a cuboidal epithelium afterward as it continues branching into smaller and finally terminal bronchioles (0.5–1.0 mm in diameter; without goblet cells). Considered the respiratory zone of the lungs (which amounts to 2.5–3 L in volume for gas exchange to take place), the terminal bronchioles continue downstream as alveolar ducts that are lined with alveoli and alveolar sacs (composed of a variable number of alveoli). As the smallest subdivision of the respiratory system, the alveoli (>300 million in total) are covered by a thin layer of cells called the *interstitium* (25 nm in thickness), formed by squamous epithelium or Type I alveolar cells (98%), which is covered by a thin film of surfactant fluid produced by septal cells or Type II alveolar cells (2%). The alveoli contain capillaries (branches from the pulmonary arteries) and openings (10–15 μm in size) that help equalize air pressure among the alveoli and facilitate the transfer of oxygen to the red blood cells and the release and transfer of carbon dioxide to the alveolar airway.

Based on the anatomic and functional differences, the lungs are often divided into central and peripheral compartments. The central compartment comprises the large and medium-sized bronchi, which are lined by pseudostratified epithelium (including basal cells, NE cells, luminal tall ciliated cells, and mucous secreting cells). The peripheral compartment comprises smaller airway bronchioles (including short, stubby ciliated cells and secretory Clara cells) and alveoli (including Type I alveolar cells and surfactant-secreting Type II alveolar cells). Although a single primordial stem cell is believed to give rise to the epithelium of the lung, the basal cells appear to be the specialized stem cells for the central compartment, and the Clara cells and Type II alveolar cells function as specialized stem cells for the peripheral airways.

Similar to other NSCLC, LCC and LCNEC likely evolve from a common pluripotent progenitor cell capable of multidirectional differentiation. Located in the outer regions of the lungs, LCC and LCNEC grow more rapidly and spread more quickly than some other forms of NSCLC. Metastases tend to occur in the hilar or mediastinal nodes followed by the pleura, liver, bone, brain, abdominal lymph nodes, and pericardium.

11.3 Epidemiology

LCC and LCNEC account for about 9% and 3% of all lung tumors, respectively. Predominantly occurring in smokers of average age at 60 years, these tumors show a male predilection (80%).

11.4 Pathogenesis

Besides association with smoking and environmental and occupational exposures, LCC contains K-ras and P53 gene mutations, RB pathway changes (loss of P16INK4, and hyperexpression of cyclin D1 or E), and high chromosomal instability (e.g., amplification of 1q21-q22, 8q and deletion of 3p12-p14, 4p, 8p22-p23, 21q). LCNEC harbors P53 and Rb mutations as well as chromosomal imbalances (e.g., allelic losses of 3p21, FHIT, 3p22-24, 5q21, 9p21; bcl2 overexpression, lack of BAX expression, and high telomerase activity) [3,4].

11.5 Clinical features

LCC and LCNEC are peripheral tumors that show similar clinical symptoms to other NSCLC. These include chronic cough and coughing up blood, fatigue, mild shortness of breath, and chest/back/shoulder ache (due to pleural effusions and tumor invasion of the chest wall). In addition, LCC and LCNEC may secrete hormone-like substances that lead to paraneoplastic syndrome (e.g., breast enlargement in men, which is referred to as *gynecomastia*).

11.6 Diagnosis

LCC and LCNEC are typically large, spherical masses located in the peripheral lung with well-defined borders and bulging, and frequent invasion of the visceral pleura, chest wall, or adjacent structures. Sectioning often reveals soft, pink-tan tumors with frequent necrosis, occasional hemorrhage, and rare cavitation.

Histologically, LCC shows sheets or nests of large polygonal cells with moderately abundant cytoplasm, round to extremely irregular nuclei, irregular chromatin distribution, and prominent nucleoli. Immunohistochemically, LCC stains positive for CK5 (56%), calretinin (38%), thrombomodulin (25%), mesothelin (13%), and TTF1 (variable) [5].

LCNEC displays organoid nesting, trabecular growth, rosetting, and perilobular palisading patterns (NE morphology); moderately large, polygonal cells (larger than atypical carcinoid) with abundant cytoplasm, coarse or vesicular nuclear chromatin, and prominent nucleoli (non-small cell cytological features); a mean of 70 mitotic rates per $2\,mm^2$ (phosphohistone H3 stain recommended); large zones of necrosis; and positivity for at least one NE marker (e.g., chromogranin, synaptophysin, CD56, or NCAM) or observation of NE granules by electron microscopy. LCNEC is positive for chromogranin, synaptophysin, CD56, GLUT1 (74%), CD117 (60%), TTF-1

(50%), and enhancer of zeste homolog 2 (EZH2, diffusely and strongly positive in all SCLC and LCNEC); but negative for CK14 and CK20 (34ßE12) [6].

LCNEM differs from LCNEC by the absence of NE markers. Combined large cell NE carcinoma is a LCNEC with adenocarcinoma, SCC, or sarcomatoid carcinoma. A tumor with small cell component is referred to as a *combined small cell carcinoma.*

Differential diagnoses for LCC include poorly differentiated squamous cell carcinoma (foci of keratinization and/or intercellular bridges), solid type adenocarcinoma (a minimum of 5 mucinous droplets in at least 2 HPF), lymphoepithelioma-like carcinoma (syncytial growth pattern with squamoid or glandular differentiation, lymphocytic infiltrate, EBV infection especially in Asian patients), clear cell carcinoma (clear to eosinophilic, finely granular cytoplasm with abundant PAS+ glycogen and stains with melanocytic and smooth muscle markers) [2].

Differential diagnoses for LCNEC include other NE tumors (i.e., typical carcinoid tumor, well-differentiated, Grade I, <2 mitoses per 2 mm^2, no necrosis; atypical carcinoid tumor, well-differentiated, Grade II, 2–10 mitoses per 2 mm^2, focal necrosis, frequent diffuse NE hyperplasia, or MEN1 mutations; SCLC, poorly differentiated, Grade III, 80 mitoses per 2 mm^2, necrosis, and less prominent nucleoli and nuclei smaller than three times the diameter of a small resting lymphocyte) (see Chapter 14, Table 14.1 and Chapter 18) [2].

Molecular testing for all non-small lung cancers should focus on the presence of mutations in the gene coding for epidermal growth factor receptor (EGFR) and possibly the ALK-EML4 translocation by FISH.

11.7 Treatment

Treatment options for LCC and LCNEC consist of surgery, chemotherapy and/or radiotherapy. Chemotherapy (e.g., Alimta [pemetrexed] and Platinol [cisplatin]) may be administered postoperatively, in conjunction with radiotherapy, or alone for LCC and LCNEC. Targeted therapy (Tarceva [erlotinib] and Iressa [gefitinib]) may be also considered for attacking lung cancer specifically, with fewer side effects than traditional chemotherapy [2,7,8].

11.8 Prognosis

The 5-year overall survival for lung cancer is about 18%. LCNEC has a poorer prognosis than LCC as it is often Stage III–IV at diagnosis and is more aggressive than LCC. Presence of allelic losses in 3p21, FHIT, 3p22-24, 5q21, 9p21, and RB gene mutation also correlates with poor prognosis in LCNEC. Further, combined large cell NE carcinoma may have an inferior survival compared to its classic counterpart.

References

1. Travis WD, Brambilla E, Burke AP, Marx A, Nicholson AG. *WHO Classification of Tumours of the Lung, Pleura, Thymus and Heart.* International Agency for Research on Cancer, Lyon; 2015.
2. Hendifar AE, Marchevsky AM, Tuli R. Neuroendocrine tumors of the lung: Current challenges and advances in the diagnosis and management of well-differentiated disease. *J Thorac Oncol.* 2017;12(3): 425–36.
3. Pelosi G, Fabbri A, Papotti M, et al. Dissecting pulmonary large-cell carcinoma by targeted next generation sequencing of several cancer genes pushes genotypic-phenotypic correlations to emerge. *J Thorac Oncol.* 2015;10(11):1560–9.
4. Rekhtman N, Pietanza MC, Hellmann MD, et al. Next-generation sequencing of pulmonary large cell neuroendocrine carcinoma reveals small cell carcinoma-like and non-small cell carcinoma-like subsets. *Clin Cancer Res.* 2016;22(14):3618–29.
5. PathologyOutlines.com website. *Large cell undifferentiated carcinoma.* http://www.pathologyoutlines.com/topic/lungtumorlargecell.html; accessed December 22, 2016.
6. PathologyOutlines.com website. *Large cell neuroendocrine carcinoma.* http://www.pathologyoutlines.com/topic/lungtumorlargecellNE. html; accessed December 22, 2016.
7. Rieber J, Schmitt J, Warth A, et al. Outcome and prognostic factors of multimodal therapy for pulmonary large-cell neuroendocrine carcinomas. *Eur J Med Res.* 2015;20:64.
8. Sala González MÁ. Prolonged survival with erlotinib followed by afatinib in a Caucasian smoker with metastatic, poorly differentiated large cell carcinoma of the lung: a case report. *Cancer Biol Ther.* 2015;16(10):1434–7.

12
Malignant Pleural Mesothelioma

Raúl Barrera-Rodríguez and Carlos Pérez-Guzmán

12.1 Definition

Tumors affecting the pleura are categorized as mesothelial, mesenchymal, or lymphoproliferative. Within the mesothelial tumor category, three types are recognized, that is, diffuse malignant mesothelioma (epithelioid, sarcomatoid, and biphasic), localized malignant mesothelioma, and other tumors of mesothelial origin (well-differentiated papillary mesothelioma and adenomatoid tumor). Among these, diffuse malignant mesothelioma (commonly known as *malignant pleural mesothelioma* or MPM, and simply as *malignant mesothelioma* or *mesothelioma*) is the most common primary malignant tumor of the pleura with a generally dismal prognosis [1].

12.2 Biology

Like other serous cavities and internal organs, the pleura is lined/covered by an extensive monolayer of pavement-like mesothelial cells, forming the mesothelium (or mesothelial membrane). In the pleura, mesothelial cells facilitate the free movement of the pleural surface during respiration by the production of a lubricating glycoprotein, and they proliferate in response to cell damage and growth factors.

After inhalation, asbestos fibers may be entrapped permanently in lung tissue. Although the majority of these fibers remain naked without causing a tissue reaction, some may irate the pleura, scratch the lung, generate reactive oxygen species, and activate the mitogen-activated protein kinase signaling pathway and its substrates, leading to their encapsulation by multinucleated macrophages, along with the deposition of protein and hemoglobin-derived iron and the subsequent formation of ferruginous bodies.

12.3 Epidemiology

Analysis of the WHO mesothelioma mortality database between 1994 and 2008 yielded an age-adjusted mortality rate of 4.9 per million for MPM, with a mean age at death of >70 years and a male-to-female ratio of 3.6:1. There appears to a marked heterogeneity in MPM incidence between countries, with 10 per million in the United States, 10–20 per million in Europe, and 60 per million men and 11 per million women in Australia [1].

12.4 Pathogenesis

About 70% of MPM cases have a history of direct or indirect exposure to asbestos (including the serpentine chrysotile and members of the amphibole family—amosite, crocidolite, tremolite, anthophyllite, and actinolite). Other agents and factors related to MPM consist of the nonasbestos fiber erionite (seen only in Cappadocia, Turkey), beryllium, therapeutic radiation, and possibly processes that lead to intense pleural scarring such as prior plombage therapy for tuberculosis. Simian virus 40 (SV40) and inactivated nuclear deubiquitinase BRCA1-associated protein 1 (BAP1) may have a role in the pathogenesis of MPM [2].

12.5 Clinical features

Patients with MPM commonly develop breathlessness (25%) with a pleural effusion frequently accompanied by nonpleuritic chest-wall pain (>60%), dyspnea (25%), cough (20%), and fatigue and weight loss (<30%). Some cases may be completely asymptomatic [3].

12.6 Diagnosis

Diagnosis of MPM involves imaging study, biopsy, pulmonary function tests, and other laboratory approaches (e.g., serology, cytology, histology, immunohistochemistry, and molecular biology) [1,4].

- **Imaging study:** Conventional chest radiography typically reveals pleural thickening, pleural effusion (usually unilateral), and occasionally pleural masses or signs of asbestosis. CT and positron emission tomography-CT (PET-CT) provide a more accurate preoperative assessment of tumor, including pleural effusion (74% of cases), basal tumor masses (92% of cases), and invasion of the chest wall (18% of cases, usually after surgery). Ultrasound and contrast-enhanced ultrasound may be useful in quantifying pleural effusion and thickening, as well as identifying discrete malignant nodules.

- **Biopsy:** Thoracoscopy is the preferred biopsy technique. Ultrasound- and CT-guided biopsies have a higher diagnostic yield (up to 90%) than blind needle biopsy of pleural thickenings or lesions. A blind needle biopsy may still be considered when the pleural cavity is inaccessible due to extensive pleural adhesion.
- **Pulmonary function tests:** Patients with MPM often show a restrictive pattern with an increase in peak expiratory flow. A change in forced vital capacity is a surprisingly accurate indicator of simple progression or regression of the disease.
- **Serology:** Serum mesothelin-related protein (SMRP) is elevated in >60% of patients with MPM and <2% of patients with other pleural or lung disease. Another biomarker, fibulin-3, identifies mesothelioma with a sensitivity of 72.9%–96.7% and a specificity of 88.5%–95.5%. Overall, SMRP gives a better diagnostic accuracy than fibulin-3 when measured in plasma or pleural fluid.
- **Cytology:** Cytological examination of effusion shows pleomorphic or bland mesothelioma cells arranged in sheets, clusters, morulae, or papillae, sometimes with psammoma bodies. Whereas benign mesothelial cells may exhibit increased cellularity, pleomorphism, and mitotic activity, malignant mesothelioma contains parakeratotic-like cells with orange cytoplasm and pyknotic nuclei on Papanicolaou-stained cytology slides. Differentiation of mesothelioma from benign mesothelial hyperplasia with reactive atypia is assisted by detection of glucose transporter 1 (Glut-1), epithelial membrane antigen, and desmin expression.
- **Histology:** MPM consists of three histological types:
  - Epithelioid mesothelioma, which accounts for 60% of cases and shows tubule papillary, glandular, and solid growth patterns, with tubulopapillary, adenomatoid (microglandular), and sheetlike most frequently observed, and small cell, clear cell, and deciduoid patterns rarely seen. MPM with anaplastic and/or tumor giant cells is designated pleomorphic.
  - Sarcomatoid mesothelioma, or *spindled* or *fibrous mesothelioma*, with desmoplastic mesothelioma as its variant, which is the rarest (8%) and most lethal mesothelioma, and shows spindle cells arranged in fascicles or having a haphazard distribution.
  - Biphasic mesothelioma, which represents about 30% of cases and contains both epithelioid and sarcomatoid components, with each component being at least 10% of the tumor.

Additional use of electron microscopy can help differentiate among sarcomatoid or desmoplastic mesothelioma, adenocarcinoma, and fibrotic pleuritis.

- **Immunohistochemistry:** The use of specific antibodies in immunohistochemical procedures enables further confirmation of MPM. The most common mesothelial markers are summarized in Table 12.1.
- **Molecular biology:** Fluorescence *in situ* hybridization (FISH) for homozygous deletion of p16/CDKN2A (which occurs in up to 80% of primary pleural mesotheliomas) may help distinguish benign from malignant mesothelioma.
- **Staging:** The stage of pleural mesothelioma is often determined by using the International Mesothelioma Interest Group TNM staging system, which incorporates the size and position of the mesothelioma primary tumor (*T*), eventual spread of mesothelioma cells to nearby lymph nodes (*N*), and eventual metastatic spread of mesothelioma cells to other parts of the body (*M*) (Table 12.2). Stage IA is defined as T1a, N0, M0; Stage IB as T1b, N0, M0; Stage II as T2, N0, M0; Stage III as T1, T2, N1, M0; or T1, T2, N2, M0; or T3, N0, N1, N2, M0; Stage IV as T4, any N, M0; or any T, N3 and M0; or any T, any N, M1 [4].

12.7 Treatment

Treatment options for MPM include surgery, chemotherapy, radiotherapy, palliation, and targeted therapies [5].

Table 12.1   Immunohistochemical Markers for Differential Diagnosis of Mesothelioma

Epithelioid Mesothelioma		Sarcomatoid Mesothelioma	Mesothelioma vs. Reactive Mesothelial Hyperplasia
Mesothelial	**Adenocarcinoma**		
Calretinin	CEA	Cytokeratin 7	Desmin
Cytokeratin 5/6	TTF-1	Cytokeratin 8/18	Pancytokeratin
HBME-1	CD15 (Leu-M1)	Cytokeratin CAM5.2	EMA, p53, GLUT-1, and IMP3 (usually negative)
Podoplanin (D2-40)	BG8	Cytokeratin AE1/AE3	
WT-1	B72.3	Cytokeratin MNF116	
Thrombomodulin	MOC31	Cytokeratin 34 BetaE12	
Mesothelin	Ber-EP4	Vimentin	
	E-cadherin		

Note: BG8, blood group 8; CEA, carcinoembryonic antigen; EMA: epithelial membrane antigen; TTF-1, thyroid transcription factor-1; WT-1, Wilms tumor.

Table 12.2   The TNM Classification

Primary Tumor (T)		
TX	Primary tumor cannot be assessed	
T0	No evidence of primary tumor	
T1	Tumor involves ipsilateral parietal pleura, with or without focal involvement of visceral pleura	
	• **T1a** Tumor involves ipsilateral parietal (mediastinal, diaphragmatic) pleura No involvement of the visceral pleura	
	• **T1b** Tumor involves ipsilateral parietal (mediastinal, diaphragmatic) pleura, with focal involvement of the visceral pleura	
T2	Tumor involves any of the ipsilateral pleural surfaces, with at least one of the following: • Confluent visceral pleural tumor (including fissure) • Invasion of diaphragmatic muscle • Invasion of lung parenchyma	
T3	Tumor involves any of the ipsilateral pleural surfaces, with at least one of the following: • Invasion of the endothoracic fascia • Invasion into mediastinal fat • Solitary focus of tumor invading the soft tissues of the chest wall • Nontransmural involvement of the pericardium	

(Continued)

Table 12.2 *(Continued)* The TNM Classification

T4	Tumor involves any of the ipsilateral pleural surfaces, with at least one of the following diffuse or multifocal invasion of soft tissues of the chest wall: <ul><li>Any involvement of rib</li><li>Invasion through the diaphragm to the peritoneum</li><li>Invasion of any mediastinal organ(s)</li><li>Direct extension to the contralateral pleura Invasion into the spine</li><li>Extension to the internal surface of the pericardium</li><li>Pericardial effusion with positive cytology</li><li>Invasion of the myocardium</li><li>Invasion of the brachial plexus</li></ul>	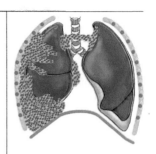
Regional lymph nodes (N) <ul><li>**NX** Regional lymph nodes cannot be assessed</li><li>**N0** No regional lymph node metastases</li><li>**N1** Metastases in the ipsilateral bronchopulmonary and/or hilar lymph node(s)</li><li>**N2** Metastases in the subcarinal lymph node(s) and/or the ipsilateral internal mammary or mediastinal lymph node(s)</li><li>**N3** Metastases in the contralateral mediastinal, internal mammary, or hilar lymph node(s) and/or the ipsilateral or contralateral supraclavicular or scalene lymph node(s)</li></ul>		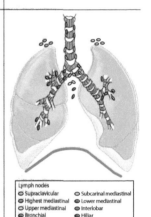

Distant metastasis (M)
- **MX** Distant metastases cannot be assessed
- **M0** No distant metastasis
- **M1** Distant metastasis

Surgical procedures can be carried out by extrapleural pneumonectomy and extended pleurectomy/decortication.

Chemotherapy alone is recommended for patients with Stage I–IV disease who are not candidates for surgery and for patients with sarcomatoid histology. This commonly involves pemetrexed (or raltitrexed) and four to six cycles of platinum , leading to a median overall survival of approximately 1 year and a median progression free survival of <6 months.

Other combinations include carboplatin and pemetrexed, cisplatin and gemcitabine.

1. First-line combination chemotherapy:
   a. Pemetrexed 500 mg/m^2 IV on day 1 plus cisplatin 75 mg/m^2; every 3 wk or
   b. Pemetrexed 500 mg/m^2 IV on day 1 plus carboplatin, every 3 wk or
   c. Gemcitabine 1000–1250 mg/m^2 IV on days 1, 8, and 15 plus cisplatin 80–100 mg/m^2 on day 1; every 3–4 wk or
   d. Amatuximab 5 mg/kg on days 1 and 8 plus pemetrexed 500 mg/m^2 and cisplatin 75 mg/m^2 on day 1 of a 21-day cycle for up to six cycles.

2. Second-line chemotherapy:
   a. Pemetrexed 500 mg/m^2 IV on day 1; every 3 wk (if not used as first-line therapy) or
   b. Vinorelbine 30 mg/m^2 IV weekly.

Radiotherapy is recommended after surgery and/or in conjunction with chemotherapy to prevent tumor spread and to reduce pain of the chest wall (with 20–60 Gy total doses, 1–7 fraction doses, over 1–6 weeks) (Table 12.3). The most effective radiotherapy technique is an intensity-modulated radiation therapy. This technique is generally used after surgical resection radical MPM, as an adequate control of local recurrence although many patients die from distant metastases. Other methods of radiation and radioactive colloids compounds and other forms of brachytherapy pleural or peritoneal cavity look attractive; however, the results are controversial.

Table 12.3   Radiation Therapies for MPM

Radiation Therapy	Total Dose	Fraction Size	Duration
Preoperative	45–50 Gy	1.8–2 Gy	4 to 5 wk
Postoperative or negative margins	50–54 Gy	1.8–2 Gy	4 to 5 wk
Microscopic-macroscopic positive margins	54–60 Gy	1.8–2 Gy	5 to 6 wk
Palliative radiation therapy or chest wall pain from recurrent nodules	20–24 Gy	4 Gy or greater	1 to 2 wk
Multiple brain or bone metastases	30 Gy	3 Gy	2 wk
Prophylactic radiation to prevent surgical tract recurrence	21 Gy	7 Gy	1 to 2 wk

Palliative measures such as removing all the liquid spill with suction pleura back, followed by an application of talc and surgical pleurodesis (bleomycin or iodine works well) will help control recurrent pleural effusion. Somatic pain (located in the chest wall) often responds to nonsteroidal analgesics, avoiding opioid addiction. Neuropathic pain (due to spinal or intercostal nerve infiltration) requires the addition of an anticonvulsant, such as *carbamazepine* or *divalproex* sodium. Opioids may be used to relieve dyspnea, such as a buildup of fluid and anaemia.

Targeted therapies rely on various new compounds to improve the outcome and survival of patients with MPM. These include tyrosine kinase inhibitors, histone deacetylase inhibitors, and immunological agents.

12.8 Prognosis

MPM often has a very poor prognosis, with median survival of 5–18 months after diagnosis. However, proper treatment can help extend survival to years. Poor prognosis factors include age (>40 years), male gender, high tumor burden, poor performance status, leukocytosis, anemia, thrombocytosis, sarcomatoid histology, high uptake values in the PET, expression of certain biochemical markers (cyclo-oxygenase-2 and VEGF), alpha $p16_{INK}4$ hypermethylated gene, increased vascularity, and evidence of SV40 virus.

References

1. Pass HI, Vogelzang NJ, Carbone M, (Editors). *Malingnant Mesothelioma: Advances in Pathogenesis, Diagnosis, and Translational Therapies.* 1st ed. New York: Springer Science+Business Media; 2005. 1–832.
2. Galateau-Sallé F, (Editor). *Pathology of Malignant Mesothelioma.* London: Springer-Verlag; 2006.
3. Pérez-Guzmán C, Barrera-Rodríguez R, Portilla-Segura J. Malignant pleural mesothelioma in a 17-year old boy: A case report and literature review. *Respir Med Case Rep.* 2016; 17: 57–60.
4. Ascoli V. Pathologic diagnosis of malignant mesothelioma: Chronological prospect and advent of recommendations and guidelines. *Ann Ist Super Sanita.* 2015; 51(1): 52–9.
5. Tai P, Joseph K, Assouline A, Au J, Yu E. Mesothelioma—Update on management. *Curr Resp Med Rev.* 2014; 10(4): 206–20.

13
Pulmonary Adenocarcinoma, Squamous Cell Carcinoma, and Adenosquamous Carcinoma

13.1 Definition

Pulmonary adenocarcinoma, squamous cell carcinoma, and adenosquamous carcinoma are non-small cell lung carcinoma (NSCLC), belonging to the epithelial tumors category according to the 2015 WHO classification of tumors of the lung, pleura, thymus, and heart [1].

Pulmonary adenocarcinoma is a very common malignant type of NSCLC that is characterized by glandular differentiation, pneumocyte phenotype, or mucin production. It accounts for 35% of all lung neoplasms.

Pulmonary squamous cell carcinoma (pulmonary SCC or squamous cell lung carcinoma) is a common type of NSCLC that is characterized by the proliferation of atypical, often pleomorphic squamous cells. The tumor is graded as well, moderately, or poorly differentiated; and well-differentiated carcinoma is usually associated with keratin production and the presence of intercellular bridges between adjacent cells. It accounts for 30% of all lung neoplasms.

Pulmonary adenosquamous carcinoma is a rare type of NSCLC that is characterized by the presence of at least 10% each of squamous and glandular differentiation components. It accounts for up to 4% of all lung neoplasms.

13.2 Biology

Pulmonary adenocarcinoma usually arises from the bronchi, bronchioles, and alveolar cells in the peripheral lung tissue, possibly through a stepwise progression from preinvasive lesions (adenocarcinoma *in situ*—mucinous, nonmucinous, or mixed; and atypical adenomatous hyperplasia), minimally invasive lesions (minimally invasive adenocarcinomas—mucinous, non-mucinous, or mixed), to invasive adenocarcinoma (acinar predominant, papillary predominant, micropapillary predominant, solid predominant with mucin production, and lepidic predominant adenocarcinoma [LPA], as well as invasive mucinous adenocarcinoma and fetal/enteric colloid) (see Chapter 17) [2]. Therefore, the peripheral adenocarcinomas often have a prominent bronchioloalveolar component. However, some adenocarcinomas may develop centrally from the surface epithelium or occasionally from the bronchial glands.

Pulmonary SCC originates from transformed central airway stem cells and generally grows in the central part of the lung (i.e., the mainstem) in close relation to the large bronchi, but occasionally in a peripheral location. The tumor often develops from a precancerous lesion proceeded by squamous metaplasia and squamous cell hyperplasia.

Pulmonary adenosquamous carcinoma is a predominantly peripheral tumor associated with scars, containing at least 10% of adenocarcinoma and squamous cell carcinoma.

13.3 Epidemiology

All histologic types of lung cancer are linked to cigarette smoking. However, it appears that pulmonary SCC, along with SCLC, is more strongly associated with smoking than pulmonary adenocarcinoma. The latter more commonly affects never-smokers (e.g., females, younger males, and smokers who have quit) than smokers (62% vs. 18%). Pulmonary adenosquamous carcinoma affects both men and women, most of whom are smokers. This tumor tends to have early metastasis and poor prognosis.

13.4 Pathogenesis

Risk factors for lung cancer include tobacco smoking, passive tobacco smoking, asbestos exposure, ionizing radiation (e.g., radon), indoor air pollution, and chronic lung disease. Over 90% of patients with pulmonary SCC are cigarette smokers, and some also have arsenic exposure.

Molecularly, pulmonary adenocarcinoma often contains epidermal growth factor receptor (EGFR) mutations, and pulmonary SCC is associated with *DDR2* mutation and FRGF1 amplification [3].

13.5 Clinical features

Patients with pulmonary adenocarcinoma, SCC, and adenosquamous carcinoma may manifest with cough, hemoptysis (blood in cough or sputum), dyspnea (shortness of breath), fatigue, weight loss, chest pain, and pleural effusion (fluid in the chest). Some patients may be asymptomatic with incidental radiologic finding of lung tumors.

13.6 Diagnosis

Apart from physical exam and medical history review, chest X-ray, CT, PET, and MRI are helpful in localizing lung cancer. Additional diagnostic approaches include bronchoscopy (to collect small lung tissue samples), thoracentesis (using a cannula to remove fluid between the lungs and the chest wall), thoracoscopy (to examine and remove tissue from the chest wall), thoracotomy (a surgical invasive procedure to remove tissue from the chest wall or the surrounding lymph nodes), mediastinoscopy (to examine and remove samples inside the chest wall) and bone marrow biopsy.

Pulmonary adenocarcinoma is a single or multiple, solid/firm, yellow-white nodule or mass showing invasion into the pleura and causing pleural retraction/puckering, and appears as a diffuse pleural thickening resembling malignant mesothelioma. Radiographically, the tumor displays ground-glass opacity and solid nodule (<3 cm) in the periphery and upper lobes and occasionally in other areas of the lung [4,5].

Histologically, lepidic predominant adenocarcinoma (LPA) is a nonmucinous adenocarcinoma composed of neoplastic cells lining the alveoli with no architectural disruption/complexity, and no lymphovascular or pleural invasion. However, observation of any mucinous component would make it an invasive adenocarcinoma. Acinar predominant adenocarcinoma is characterized by glandular formation. Papillary predominant adenocarcinoma shows true fibrovascular cores lined by tumor cells replacing the alveolar lining. Micropapillary predominant adenocarcinoma is made up of ill-defined projection/tufting with no fibrovascular cores, and psammoma bodies may be present in the papillary and micropapillary variants. Adenocarcinoma with histologic patterns other than lepidic, infiltration with desmoplastic reaction, lymphovascular or pleural invasion, or necrosis is classified as invasive tumor. Variants include invasive mucinous adenocarcinoma (formerly mucinous BAC), colloid, fetal, and enteric adenocarcinoma. Adenocarcinoma of <3 cm in size or with pure lepidic pattern but no features of invasion is referred to as adenocarcinoma *in situ* (AIS), and bronchioloalveolar carcinoma (BAC) is no longer used. Solitary tumor of <3 cm or with predominantly lepidic pattern and ≤5 mm invasion in any one focus is considered minimally invasive adenocarcinoma (MIA) [1,4,5].

Immunohistochemically, pulmonary adenocarcinoma stains positive for CK7, TTF1 (sensitivity 84.5%, specificity 96.4%), NapsinA (sensitivity 92.0%, specificity 100%), mucin, epithelial membrane antigen (EMA), carcinoembryonic antigen (CEA), surfactant apoprotein (50%), mesothelin (50%), vimentin (9%), S100 (Langerhans cells), p53, CD57/Leu7 (50%), calretinin (11%), and EGFR mutation-specific antibodies (variable) but negative for CK20, keratin 5 (usually), P504S, and p63 (cytoplasmic expression associated with bad prognosis).

At the molecular/cytogenetic level, pulmonary adenocarcinoma often harbors mutations related to *EGFR* (10%–15%, which responds to treatment with tyrosine kinase inhibitors), *KRAS* (15%–25%), and *BRAF*; fusion between echinoderm microtubule-associated protein-like 4 (*EML4*) and *ALK* (2%–7%), which benefits from treatment with ALK inhibitors, and amplification of MET (a heterodimer receptor tyrosine kinase involved in organogenesis) and *FGFR1*. Interestingly, *EGFR*, *KRAS*, and *ALK* mutations appear to be mutually exclusive.

For differential diagnosis, pulmonary adenocarcinoma is negative for p63 and p40 (SCC is positive); synaptophysin, chromogranin, and CD56 (neuroendocrine tumors are positive); calretinin, D240, CK5/6, and WT1 (mesothelioma is positive) [4,5].

Pulmonary SCC usually occurs in the central portion of the lung affecting larger bronchi but occasionally in peripheral location, and often invades the peribronchial soft tissue, lung parenchyma, and nearby lymph nodes and compresses the pulmonary artery and vein. Peripheral tumor tends to show nodular growth with central necrosis and cavitation (in contrast, adenocarcinoma usually does not form a cavitary lesion), and surrounding lung may exhibit lipid pneumonia, bronchopneumonia, and atelectasis [6].

Histologically, pulmonary SCC displays sheets or islands of large polygonal malignant cells containing keratin (individual cells or keratin pearls) and intercellular bridges; adjacent bronchial dysplasia or carcinoma *in situ*; destruction of alveoli or filling of alveolar spaces at advancing tumor border, and rare spreads beneath basement membrane; focal areas of intracytoplasmic mucin; occasional oncocytes, foreign body giant cells (reacting to keratin), palisading granulomas, extensive neutrophilic infiltration, lepidic growth pattern at tumor periphery, clear cell change (glycogen); and alveolar space filling (tumor cells fill alveoli but don't destroy elastic septa), expanding type (growth destroys elastic septa), or mixture in peripheral tumor types. The tumor may be classified as well, moderately, or poorly differentiated based on the amount of keratinization present in the predominant component. The tumor stains positive for p63, CK5/6 (87%–100%), EMA, thrombomodulin (87%–100%); variably positive for CD15, CEA, HPV, mesothelin (16%–31%), p53, p40, S100; but negative for vimentin (usually), TTF1 (usually), and Napsin A [7,8].

Pulmonary adenosquamous carcinoma usually presents as a single mass of 2.8–3.8 cm (range 0.6–11 cm) in size on the periphery of the lung. The tumor stains positive for TTF1 in the adenocarcinoma component and p63 in the squamous component.

13.7 Treatment

Treatment options for adenocarcinoma, squamous cell carcinoma, and adenosquamous carcinoma of the lung include surgery, radiation therapy, and/or chemotherapy [9].

Surgery involves wedge resection (partial removal of the wedge-shape portion of the lung that contains cancerous cells along with any surrounding healthy tissue), segmentectomy (partial removal of the cancerous lung and any surrounding healthy tissue), lobectomy (a most common procedure to partially remove a portion of a lung), sleeve lobectomy (a surgical procedure to partially remove a portion of the lung and a part of the airway or bronchus), or pneumonectomy (surgical removal of the entire lung. Surgery is particularly useful for early stage disease, with adenocarcinoma in situ (AIS) and minimally invasive adenocarcinoma (MIA) often having 100% or near 100% disease-specific survival, respectively after complete resection.

Radiation therapy can be applied either externally (external beam radiation) or internally (brachytherapy).

Chemotherapy utilizes drugs (e.g., erlotinib or Tarceva) to kill and inhibit cancer cells. However, treatment of squamous cell lung carcinoma with bevacizumab may sometimes lead to life-threatening hemorrhage

13.8 Prognosis

With a size of >2.5 cm, micropapillary and solid variants of pulmonary adenocarcinoma have a poor prognosis, whereas lepidic pattern has a much better prognosis. Adenocarcinoma with KRAS mutation and MET amplification are associated with poor prognosis and EGFR-acquired resistance. The 5-year survival rates for adenocarcinoma are 49% (Stage IA), 45% (Stage IB), 30% (Stage IIA), 31% (Stage IIB), 14% (Stage IIIA), 5% (Stage IIIB), and 1% (Stage IV).

Adenosquamous carcinoma of the lung has a poorer prognosis than adeno-carcinoma or squamous cell carcinoma of the lung. Genetic mutation in Ras or p53 is also associated with a poorer prognosis.

References

1. Travis WD, Brambilla E, Burke AP, Marx A, Nicholson AG. *WHO Classification of Tumours of the Lung, Pleura, Thymus and Heart*. International Agency for Research on Cancer, Lyon; 2015.

2. Gardiner N, Jogai S, Wallis A. The revised lung adenocarcinoma classification—An imaging guide. *J Thorac Dis*. 2014; 6(Suppl 5): S537–46.

3. Gandara DR, Hammerman PS, Sos ML, Lara PN Jr, Hirsch FR. Squamous cell lung cancer: From tumor genomics to cancer therapeutics. *Clin Cancer Res*. 2015;21(10):2236–43.

4. PathologyOutlines.com website. *Adenocarcinoma-classification*. http://www.pathologyoutlines.com/topic/lungtumoradenoclass.html; accessed December 22, 2016.

5. Roviello G. The distinctive nature of adenocarcinoma of the lung. *Onco Targets Ther*. 2015;8:2399–406.

6. Gerber DE, Paik PK, Dowlati A. Beyond adenocarcinoma: Current treatments and future directions for squamous, small cell, and rare lung cancer histologies. *Am Soc Clin Oncol Educ Book*. 2015:147–62.

7. PathologyOutlines.com website. *Squamous cell carcinoma*. http://www.pathologyoutlines.com/topic/lungtumorSCC.html; accessed December 22, 2016.

8. Krimsky W, Muganlinskaya N, Sarkar S, et al. The changing anatomic position of squamous cell carcinoma of the lung—a new conundrum. *J Community Hosp Intern Med Perspect*. 2016;6(6):33299.

9. Derman BA, Mileham KF, Bonomi PD, Batus M, Fidler MJ. Treatment of advanced squamous cell carcinoma of the lung: A review. *Transl Lung Cancer Res*. 2015;4(5):524–32.

14

Pulmonary Carcinoid Tumor

14.1 Definition

Pulmonary carcinoid tumor (or *bronchopulmonary carcinoid*) is an unusual, non-small cell lung cancer (NSCLC) with neuroendocrine (NE) morphology and differentiation, and thus belong to a group of tumors known as NE tumors (NET).

Pulmonary carcinoid tumor is divided into two subtypes: typical carcinoid and atypical carcinoid. Typical carcinoid tumor (TC) is a more common and well-differentiated subtype that exhibits a distinct histologic pattern (insular, trabecular, glandular, mixed, or undifferentiated) of carcinoid morphology and demonstrates <2 mitoses/2 mm^2 and absence of necrosis. Atypical carcinoid tumor (AT) is a less common, moderately differentiated subtype that exhibits increased nuclear atypia, 2–10 mitoses/2 mm^2, and foci of punctate necrosis [1,2].

In comparison with other NET in the lung, that is, large cell NE carcinoma (LCNEC) and small cell lung carcinoma (SCLC), which are poorly differentiated, Grade III tumors with >10 mitoses/2 mm^2 (usually >50 mitoses/2 mm^2) and extensive geographic necrosis (see Chapters 11 and 18), TC and AC are well-differentiated and moderately differentiated NET of Grades I and II, respectively (Table 14.1) [2].

14.2 Biology

Pulmonary NE cells (PNEC) are specialized epithelial cells of the airways and lungs that form solitary cells or clusters of 4–10 cells (neuroepithelial bodies). Functioning as chemoreceptors in the airway, neuroepithelial bodies secrete serotonin in response to hypoxia, induce vasoconstriction in poorly ventilated areas of the lung, and redirect blood flow toward better-ventilated areas.

Carcinoid tumor (TC and AC) of the lung is thought to arise *de novo* from Kulchitsky cells (or *enterochromaffin cells*) disseminated throughout the bronchopulmonary mucosa or evolve from a preinvasive lesion known as *diffuse idiopathic NE cell hyperplasia* (DIPNECH) or hyperplasia of PNEC, which is an exaggerated adaptive response to airway fibrosis, hypoxia and chronic lung injury or infection, leading to chronic nonproductive cough,

Table 14.1 Key Features of Pulmonary Neuroendocrine Tumors

Tumor	Percentage of all Lung Tumors	Grade	Differentiation	Mitoses/2 mm²	Necrosis	Nuclear Pleomorphism/ Hyperchromatism	Nucleoli	Nuclear/ Cytoplasmic Ratio	Chromosome 11q Deletion
Typical carcinoid (TC)	2	I	Well-differentiated	<2	Absent	Uncommon	Occasional	Moderate	47%
Atypical carcinoid (AC)	0.2	II	Moderately differentiated	2–10	Present (focal, punctate)	Sometimes	Common	Moderate	66%
Large cell neuroendocrine carcinoma (LCNEC)	3	III	Poorly differentiated	70 (mean)	Present (large zone)	Frequent	Very common	Low	Rare
Small cell lung carcinoma (SCLC)	15–20	III	Poorly differentiated	80 (mean)	Present (large zone)	Small cells	Absent or inconspicuous	High	Rare

interstitial lung disease, and finally obliterative bronchiolitis. Normally, DIPNECH is confined to the respiratory epithelium without penetration through the basement membrane. When DIPNECH lesion extends beyond the basement membrane, it is known as *tumorlet* (with nodular aggregates of NE cells measuring ≤0.5 cm in diameter) or as *lung NET* (with nodular aggregates measuring ≥0.5 cm) (see Chapter 17) [2].

Approximately 80% of carcinoid tumor (mainly TC) develops within a bronchus of subsegmental size or greater located centrally (or prehilarly), of which 10%–15% arise in a mainstem bronchus; the remaining 20% of carcinoid tumor (mainly AT) are found in the peripheral lung [3].

14.3 Epidemiology

TC and AC account for about 2% and 0.2% of primary lung neoplasms, respectively, with a TC–AC ratio of about 9:1. Pulmonary carcinoid occurs mainly in nonsmokers or current light smokers, with more AC patients being current or former smokers than TC patients. Recent data from the SEER Registry indicate that annual incidence rates of bronchopulmonary carcinoids are 0.52 and 0.89 per 100,000 in white males and females, respectively, and 0.39 and 0.57 per 100,000 in black males and females, respectively. The mean age at diagnosis is 45 years for TC and 55 years for AC [3].

14.4 Pathogenesis

Carcinoid tumor (TC and AC) is clearly linked to cigarette smoking. Molecularly, several gene mutations are implicated in the tumorigenesis of carcinoid tumor, including multiple endocrine neoplasia Type 1 gene (MEN1) at 11q13 and *SMAD4* (TC); *KIT*, *PTEN*, *HNF1A*, and *SMO* (AC), loss of Rb expression, and loss of chromosome 3p. However, p53 mutations and chromosomal loss of 5q21 are associated with more aggressive NE tumors.

Compared to the normal cells, the cells in most carcinoid tumors contain membrane-bound secretory granules with dense-core granules in the cytoplasm. These granules produce a diversity of biogenic amines (e.g., 5-hydroxytryptamine [5-HT], 5-hydroxyindoleacetic acid, 5-hydroxytryptophan, dopamine, and histamine), peptides (atrial natriuretic peptide, chromogranins A/C, α1-antitrypsin, neurotensin, vasoactive intestinal polypeptide, pancreatic polypeptide, motilin, human chorionic gonadotropin α/β, somatostatin, and substance P), tachykinins (kallikrein and neuropeptide K), and others (prostaglandins). Among these NE tumor-related molecules, 5-HT (serotonin) is a vasoactive peptide synthesized from the amino acid tryptophan. As the most common biologically active substance secreted from carcinoid tumor, 5-HT is capable of inducing classic symptoms of carcinoid syndrome, such as diarrhea, episodic flushing, bronchoconstriction, and right-sided valvular heart disease after gaining entry to the bloodstream [4].

14.5 Clinical features

Patients with carcinoid tumor (especially those with centrally located tumor) may present with cough, hemoptysis, recurrent pulmonary infection, dyspnea, and chest pain (similar to those observed in asthma, chronic obstructive pulmonary disease, and pneumonia); and up to half of patients (especially those with peripheral tumors) are asymptomatic.

14.6 Diagnosis

For suspected pulmonary NET, imaging is performed to determine tumor size and location and for staging. While CT is recommended for imaging lung carcinoid tumor, MRI is useful for imaging mediastinal or abdominal metastasis. As somatostatin (SST) receptors (especially SST_2) are expressed in 80% of lung NET, SST receptor scintigraphy (SRS) with radiolabeled SSA 111In-labeled octreotide, in conjunction with single-photon emission CT, is effective for visualizing TC and AC.

Pulmonary carcinoid tumor is either a central (polypoid and endobronchial in major bronchi) or peripheral (solid/nodular) mass of ≥0.5 cm with a well-defined, smooth, ivory to pink cut surface. Nearly 60% of pulmonary carcinoid tumors are located in the right lung.

Definitive diagnosis of carcinoid tumor currently relies on histology (morphology, mitotic account, and presence or absence of necrosis) followed by confirmation with positive immunohistochemical staining (i.e., positive for one or more NE markers such as chromogranin A or synaptophysin) or electron microscopy (showing membrane-bound secretory granules with dense-core granules in the cytoplasm).

Microscopically, carcinoid tumor contains nests or trabeculae of medium-sized polygonal cells with lightly eosinophilic cytoplasm, low nuclear-grade, round to oval, finely granular nuclei; possible rosettes or small acinar structures with variable mucin; scanty vascular stroma, and occasionally amyloid stroma with bone. Despite their near-identical histomorphology, TC shows <2 mitoses per 10 high power field (HPF) (Ki-67 staining) and no necrosis, whereas AC has 2–10 mitoses per 10 HPF or focal necrosis (punctate foci within tumor nests) [2,5].

Carcinoid tumor stains positive for NE markers (e.g., synaptophysin, chromogranin, CD56, and neuron-specific enolase) and variable for thyroid transcription factor-1 (TTF1) (43%–53%) and secretin receptor. AC is also positive for pancreatic polypeptide, foxM1, p27/kip1 (high), and PAX5 with c-Met but negative for enhancer of zeste homolog 2 and collapsin response mediator protein (CRMP5) [6,7].

Staging of typical and atypical carcinoids follows the TNM guidelines. Overall, the proportions of TC diagnosed at Stages I, II, III, and IV are 87%, 10%, 3%, and 2%, respectively; and those of AT diagnosed at Stages I, II, III, and IV are 43%, 29%, 28%, and 21%, respectively [8].

Differential diagnoses include SCLC and LCNEC (which have higher mitotic rates and proliferation rates and more extensive necrosis; SCLC also has less cytoplasm and thus appears more hyperchromatic than AC) (Table 14.1).

14.7 Treatment

Treatment for patients with small (<2 cm), localized pulmonary carcinoid tumor and adequate functional pulmonary reserve is conservative resection via a wedge or segmental resection, which leads to low recurrence rates and excellent long-term survival [5,9].

Patients with large (>2 cm) pulmonary carcinoid tumor showing extensive central bronchopulmonary or peripheral parenchymal involvement may require more extensive surgical resection such as lobectomy or pneumonectomy, together with systematic radical mediastinal lymphadenectomy [5,9].

Chemotherapy and radiotherapy are relatively ineffective for TC and AC. Lanreotide and octreotide may be utilized to reduce excessive hormone production and control symptoms associated with carcinoid syndrome. Interferon-α (2a or 2b) may also be used either alone or with somatostatin analogs (octreotide or lanreotide) to treat carcinoid syndrome [10]. Other agents under evaluation for TC and AC include fluorouracil, fluorouracil with streptozocin, dacarbazine, epi-adriamycin, oxaliplatin plus capecitabine, temozolomide, and everolimus plus octreotide long-acting repeatable (LAR).

14.8 Prognosis

Pulmonary carcinoid tumor generally demonstrates an indolent disease course. However, AC appears to be more aggressive than TC, with a higher frequency of mediastinal lymph node involvement (50%–60% vs. 10%–15%)

and distant metastasis (20% vs. 2%–5%), a more frequent postoperative recurrence (25% vs. 3%–5%), and worse survival outcomes at 5 years (70% vs. 90%) and 10 years (50% vs. 80%).

References

1. Travis WD, Brambilla E, Burke AP, Marx A, Nicholson AG. *WHO Classification of Tumours of the Lung, Pleura, Thymus and Heart.* International Agency for Research on Cancer, Lyon; 2015.
2. Hendifar AE, Marchevsky AM, Tuli R. Neuroendocrine tumors of the lung: Current challenges and advances in the diagnosis and management of well-differentiated disease. *J Thorac Oncol.* 2016. pii: S1556-0864(16)33515-8.
3. Wolin EM. Challenges in the diagnosis and management of well-differentiated neuroendocrine tumors of the lung (typical and atypical carcinoid): Current status and future considerations. *Oncologist.* 2015; 20(10): 1123–31.
4. Pinchot SN, Holen K, Sippel RS, Chen H. Carcinoid tumors. *Oncologist.* 2008; 13(12): 1255–69.
5. Kaifi JT, Kayser G, Ruf J, Passlick B. The diagnosis and treatment of bronchopulmonary carcinoid. *Dtsch Arztebl Int.* 2015; 112(27–28): 479–85.
6. PathologyOutlines.com website. *Carcinoid tumor.* http://www.pathologyoutlines.com/topic/lungtumorcarcinoid.html; accessed December 30, 2016
7. PathologyOutlines.com website. *Atypical carcinoid tumor.* http://www.pathologyoutlines.com/topic/lungtumoratypicalcarcinoid.html; accessed December 30, 2016
8. Bertino EM, Confer PD, Colonna JE, Ross P, Otterson GA. Pulmonary neuroendocrine/carcinoid tumors: A review article. *Cancer.* 2009; 115(19): 4434–41.
9. Noel-Savina E, Descourt R. Focus on treatment of lung carcinoid tumor. *Onco Targets Ther.* 2013; 6: 1533–7.
10. Caplin ME, Baudin E, Ferolla P, et al. Pulmonary neuroendocrine (carcinoid) tumors: European Neuroendocrine Tumor Society expert consensus and recommendations for best practice for typical and atypical pulmonary carcinoids. *Ann Oncol.* 2015; 26(8): 1604–20.

15
Pulmonary Mesenchymal Tumor

15.1 Definition

Mesenchymal tumors encompass a diverse group of neoplasms that grow from any tissues derivative of the mesenchyma (e.g., conjunctive, adipose, muscular, vascular, osteous, cartilage tissues, synovial membranes, and serous tunics) without organ specificity.

According to the 2015 *WHO Classification of Tumours of the Lung, Pleura, Thymus and Heart*, mesenchymal tumors arising from the lung (i.e., pulmonary mesenchymal tumors) include pulmonary hamartoma, chondroma, PEComatous tumors (lymphangioleiomyomatosis, PEComa [benign/clear cell tumor]; PEComa malignant), congenital peribronchial myofibroblastic tumor, diffuse pulmonary lymphangiomatosis, inflammatory myofibroblastic tumor, epithelioid hemangioendothelioma, pleuropulmonary blastoma, synovial sarcoma, pulmonary artery intimal sarcoma, pulmonary myxoid sarcoma with *EWSR1–CREB1* translocation, and myoepithelial tumors (myoepithelioma, myoepithelial carcinoma) [1].

Among these, pulmonary hamartoma (PH) and lymphangioleiomyomatosis (LAM) within the PEComatous tumor group are relatively common; whereas others are rare, with 200 or fewer cases each reported to date (including 19 cases of congenital peribronchial myofibroblastic tumor [CPMT], 36 cases of diffuse pulmonary lymphangiomatosis, about 125 cases of epithelioid hemangioendothelioma, about 120 cases of pleuropulmonary blastoma, 200 cases of pulmonary artery intimal sarcoma, and 15 cases of pulmonary myxoid sarcoma).

15.2 Biology

PH is a benign neoplasm that contains varying amounts of at least two mesenchymal elements (e.g., cartilage, fat, connective tissue, smooth muscle, and bone) together with respiratory epithelium entrapped by expanding mesenchymal growth. Apart from pulmonary hamartoma, hamartomas formed in other parts of the body are generally not regarded as true neoplasms.

PEComatous tumors are thought to arise from perivascular epithelioid cells (PECs), which may be round to spindled in shape, with variably clear to granular and eosinophilic cytoplasm. PEComatous tumors of the lung include (i) LAM (a low-grade destructive metastasizing neoplasm characterized by the proliferation of abnormal, plump spindle-shaped, myoid/smooth muscle-like cells or LAM cells with typically pale eosinophilic cytoplasm, leading to the formation of thin-walled cysts in the lungs and cystic structures in the axial lymphatics), (ii) PEComa, benign including clear cell tumor (clear cell tumor is a rare, benign/borderline peripheral lung neoplasm, consisting of sheets of rounded or oval cells with clear or eosinophilic cytoplasm and PAS+ glycogen granules), and (iii) PEComa, malignant (which is a diffuse proliferation with overlapping features between LAM and clear cell tumor) [2].

CPMT is thought to develop during the early weeks of intrauterine life from pluripotent mesenchymal cells with potential to differentiate into cartilage or myofibroblasts.

15.3 Epidemiology

PH constitutes ~75% of all benign lung tumors and ~3% of all lung tumors, with an incidence of 0.25% in the general population. PH usually affects people in the fourth and fifth decades of life and rarely children. A male predilection (2.5:1) is noted.

LAM represents <1% of cases of diffuse parenchymal lung disease, with a predominant occurrence in women of reproductive age (mean age at diagnosis of 40 years). LAM may develop sporadically or in association with tuberous sclerosis complex (TSC), which affects about 1.5 to 2.0 million people worldwide and occurs in about 1 of every 6,000 births. It is estimated that approximately 250,000 patients suffer from TSC-LAM.

CPMT is a rare primary lung mesenchymal neoplasm of neonates, showing no predilection for sex, lobe, or laterality.

15.4 Pathogenesis

PH has been found to harbor recombinations between chromosomal bands 6p21 and 14q24 and the translocation t(3;12)(q27-28;q14-15), which results in gene fusion of the high-mobility group protein gene *HMGA2* (exons 1–3) and the *LPP* gene (exons 9–11).

LAM is a low-grade, destructive, metastasizing neoplasm due to the abnormal growth of smooth muscle cells that leads to blockage of the bronchial tubes and lymphatic vessels and formation of holes or cysts in the lung. LAM may occur sporadically (S-LAM), and in association with TSC (TSC-LAM).

TSC is an autosomal dominant syndrome characterized by hamartoma formation in multiple organ systems, cerebral calcifications, seizures, and cognitive defects. Up to 40% of LAM-affected women suffer from TSC, with LAM cells from these patients usually containing growth-promoting biallelic mutations in the tuberous sclerosis genes *TSC1* (40%) and *TSC2* (60%). The *TSC1* gene encodes a 130-kDa protein (hamartin) that is implicated in the reorganization of the actin cytoskeleton. The *TSC2* gene encodes a 198-kDa protein (tuberin) that is involved in cell growth and proliferation and interaction with hamartin through its N-terminus. The hamartin–tuberin complex inhibits a kinase known as the *mammalian target of rapamycin* (mTOR), a central regulator of cell growth. It is thought that an initial mutation in either *TSC1* or *TSC2* followed by loss of heterozygosity leads to the loss of function of either *TSC1* or *TSC2* gene products. In S-LAM, two acquired mutations (usually in *TSC2*) are often found. In TSC-LAM, one germline mutation (usually in *TSC2*) and one acquired mutation are detected [3,4].

Clear cell tumor occurs infrequently in patients with tuberous sclerosis complex (with TSC2 loss of heterozygosity), LAM, or micronodular pneumocyte hyperplasia (MNPH). Clear cell tumor with diffuse features that overlap with LAM is known as *diffuse PEComatosis*.

CPMT is usually associated with non-immune hydrops fetalis and polyhydramnios in addition to complex rearrangement involving chromosomes 4, 8 and 10.

15.5 Clinical features

PH is often asymptomatic, although some patients may have hemoptysis, bronchial obstruction, and cough (especially endobronchial types).

LAM may manifest with dyspnea (shortness of breath), chest pain, cough (sometimes with phlegm or blood streaks, i.e., hemoptysis), wheezing, pneumothorax (collapsed lung), pleural effusions, lymphatic obstruction (chylothorax, chyloperitoneum, chylopericardium, and chyluria), and renal angiomyolipoma.

CPMT is associated with respiratory distress, heart failure, polyhydramnios, fetal hydrops, intrauterine fetal demise, and/or premature delivery.

15.6 Diagnosis

PH is typically a slow-growing, solitary, well-circumscribed, unencapsulated nodule (about 2.5–4 cm, occasionally >10 cm in diameter) with smooth or lobulated margins occurring in the periphery of the lung (>90%), and endobronchial hamartoma accounts for only 5% of cases. On CT, the tumor may show fat (60%, localized or generalized within the nodule) and calcification/ossification (classically popcorn type, 5%–50%) but no cavitation. On MRI, the tumor gives a heterogeneous signal (with foci of high signal representing fat, and low-signal regions representing fibrous or calcific material) on T1; high signal due to fat and cartilaginous components, and low-signal regions representing fibrous or calcific material on T2; and heterogeneous enhancement on T1 C+ (Gd). Histologically, pulmonary hamartoma shows disorganized, mature tissue elements (smooth muscle, epithelial and stromal elements), with occasional fatty tissue (which is called *lipomatous hamartoma* or *endobronchial lipoma*).

LAM usually forms well-circumscribed, round, and thin-walled cysts of 0.2–2 cm (rarely 10 cm) in diameter, which are scattered in a bilateral symmetric pattern, without any lobar predominance. In addition to diffuse bilateral reticular infiltrates (80% of cases), LAM may be associated with pneumothorax (30%–40% of cases), pleural effusion (10%–20% of cases), and hyperinflation. Histologically, LAM lesions are characterized by lung nodules or small cell clusters of LAM cells near cysts and along pulmonary bronchioles, blood vessels, and lymphatics. Two types of LAM cell subpopulations exist: myofibroblast-like spindle-shaped cells expressing smooth muscle-specific proteins (e.g., α-actin, desmin, and vimentin), and epithelioid-like cells expressing glycoprotein gp100 (a marker of melanoma cells and immature melanocytes showing immunoreactivity with human melanoma black 45 [HMB45] monoclonal antibody). LAM cells often possess eosinophilic cytoplasm with perinuclear vacuoles and oval to cigar-shaped nuclei with fine to vesicular chromatin. Infiltration of LAM cells into the blood vessels results in accumulation of hemosiderophages. The pleura may show reactive eosinophilic pleuritis in cases with pneumothorax. Patients with TSC may display MNPH, clear cell tumor, angiomyolipoma, and localized angiomyolipoma-like infiltrative lesions. Immunohistochemically, LAM cells are positive for smooth muscle actin (SMA), desmin, vimentin, estrogen receptor, progesterone receptor, β-catenin, cathepsin-κ, HMB45, microphthalmia transcription factor (MITF), and MART-1 (melan-A) (focally for the epithelioid LAM cells) [4].

Clear cell tumor is a solitary, round, circumscribed, glistening, red-tan to brown mass of <5 cm with uniform cut surface and foci of necrosis. Histologically, the tumor contains thick cords and rounded nests with sinusoid-type vascular network or sheets of epithelioid cells, clear to eosinophilic granular cytoplasm, well-defined cell borders, PAS+ glycogen granules, oval nuclei, small nucleoli, and prominent sclerotic vasculature. It rarely contains malignant features. In aggressive form (diffuse PEComatosis), the tumor may show nuclear pleomorphism, >5 cm size, increased mitoses, infiltrative growth pattern with characteristic thin-walled sinusoidal vessels, and necrosis. The tumor stains consistently positive for HMB45, MART1/melan A, MITF, SMA, desmin, CD1a, and S100 (variable) but negative for keratins and epithelial membrane antigen [5].

CPMT is a well-circumscribed, non-encapsulated mass of a few centimeters in dimension, with solid and grey-tan cut surface and necrotic areas at the center. Histologically, the uniformly spindle-shaped tumor cells contain finely granular chromatin and surround the bronchial walls. The tumor shows central necrosis, 4–5 mitoses per 10 HPF (but absence of atypical mitoses), and multifocality. The tumor is strongly positive for vimentin, but negative for SMA, desmin, S–100, CD34, human herpesvirus-8 (HHV-8), activin receptor-like kinase-1 (ALK-1), AFP, and NSE [6].

15.7 Treatment

Peripheral, small PH with no atypical features can be safely left alone. Those with atypical features or endobronchial hamartomas causing distal complications are curative by surgical resection (e.g., rigid transbronchial resection).

Currently, no effective treatment is available for LAM, which affects predominantly premenopausal females. Use of sirolimus (also known as *rapamycin,* a macrolide compound with immune suppression function) helps stabilize the disease for the majority of patients. Progesterone may be considered for individual cases with rapid progression of the disease.

Early surgical excision (lobectomy or pneumonectomy) is advisable for CPMT, but chemotherapy and/or radiotherapy are not recommended.

15.8 Prognosis

PH is a benign tumor with excellent prognosis, as it does not have the tendency to transform into a malignancy, nor recur after complete removal.

LAM is a progressive disease, with the propensity to get worse over time without treatment. Prognosis for patients with LAM was considered poor previously, but with the introduction of sirolimus, >90% of patients are alive 10 years after diagnosis.

CPMT has a good prognosis after surgery, and post-operative recurrence or metastasis has not been reported.

References

1. Travis WD, Brambilla E, Burke AP, Marx A, Nicholson AG. *WHO Classification of Tumours of the Lung, Pleura, Thymus and Heart.* International Agency for Research on Cancer, Lyon; 2015.
2. Travis WD, Brambilla E, Nicholson AG, et al. The 2015 World Health Organization classification of lung tumors: Impact of genetic, clinical and radiologic advances since the 2004 classification. *J Thorac Oncol.* 2015; 10(9): 1243–60.
3. Juvet SC, McCormack FX, Kwiatkowski DJ, Downey GP. Molecular pathogenesis of lymphangioleiomyomatosis: Lessons learned from orphans. *Am J Respir Cell Mol Biol.* 2007; 36(4): 398–408.
4. Harari S, Torre O, Moss J. Lymphangioleiomyomatosis: What do we know and what are we looking for? *Eur Respir Rev.* 2011; 20(119): 34–44.
5. PathologyOutlines.com website. Clear cell (sugar) tumor. http://www.pathologyoutlines.com/topic/lungtumorclearcell.html; accessed January 5, 2017.
6. Xia B, Yu G, Hong C, Zhang L, Tang J, Liu C. A congenital peribronchial myofibroblastic tumor detected in a premature infant at 28 weeks but that resolved in the late stage of pregnancy: A case report. *Medicine (Baltimore).* 2015; 94(42): e1842.

16

Pulmonary Sarcomatoid Carcinoma

16.1 Definition

Pulmonary sarcomatoid carcinoma (PSC, or *sarcomatoid carcinoma of the lung*) is a heterogeneous, poorly differentiated non-small cell lung carcinoma (NSCLC) showing sarcomatous or sarcomatoid morphology with giant and/or spindle cells. PSC has been differentiated into five subtypes: pleomorphic carcinoma, spindle cell carcinoma, giant cell carcinoma, carcinosarcoma, and pulmonary blastoma [1,2].

16.2 Biology

PSC is thought to evolve from the embryonic rest, proliferation of malignant epithelium and mesenchyme, stromal induction/metaplasia or a single stem cell (totipotential). However, evidence from immunohistochemical, ultrastructural, and molecular studies lends support for the totipotential origin of sarcomatoid carcinoma. The tumor can arise centrally or peripherally but most commonly occurs as a solitary, peripheral mass in the upper lobes of the lung.

Although sarcomatoid carcinoma occurs throughout the body, primary sarcomatoid carcinoma in the lung is extremely rare. PSC generally runs an aggressive clinical course with advanced local disease and metastasis. Because of its heterogeneity, PSC often poses significant challenge in diagnosis.

16.3 Epidemiology

PSC is an uncommon neoplasm, accounting for about 1.3% of all lung malignancies. The tumor is four to seven times more commonly seen in men than in women, with a mean age of 65 years at diagnosis. However, the pulmonary blastoma subtype affects men and women equally in the fourth decade [3].

16.4 Pathogenesis

PSC is linked to smoking and asbestos exposure (implicated in 3% of cases). Molecular abnormalities relating to the pathogenesis of PSC include overexpression of epidermal growth factor receptor (EGFR) protein, a high level of polysomy (four or more copies of the *EGFR* gene), and *KRAS* mutations (e.g., Gly12Cys and Gly12Val) [4].

16.5 Clinical features

Patients with PSC typically have chest pain, cough, hemoptysis, dyspnea, fever, weight loss, and weakness.

In the event of gastrointestinal metastases, clinical symptoms include abdominal pain, bleeding, obstruction, perforation, and small intestinal intussusception (as demonstrated in cases involving metastatic lung carcinosarcoma) [5].

16.6 Diagnosis

PSC is a relatively large, soft/fleshy, or firm/hard/rubbery mass of 2–17 cm (mean, 8 cm), with central (endobronchial) or peripheral (parenchymal) location in the upper lobes of the lung and frequent invasion of the pleural and/or chest wall. It has white-gray to tan-yellow cut surfaces and shows areas of hemorrhage and necrosis and occasional cavitation. Immunohistochemically, PSC is strongly positive for cytokeratin (92.9%) and vimentin (90.5%) and moderately positive for epithelial membrane antigen (EMA) (90.0%), TTF-1 (33.3%), P63 (43.8%), CK5/6 (42.3%), and CK7 (65.4%), and surfactant protein A (39% for epithelial component, 6% for sarcomatoid component) [6].

Among PSC subtypes, *pleomorphic carcinoma* (accounting for 24% of PSC cases) is a poorly differentiated, mixed carcinoma composed of at least 10% malignant spindle cells and/or giant cells plus epithelial components (e.g., adenocarcinoma, large cell carcinoma, and squamous cell carcinoma) or only of malignant spindle or giant cells (which are referred to as *spindle cell carcinoma of the lung* [11% of PSC cases] or *giant cell carcinoma of the lung* [6.5% of PSC cases]). The neoplastic spindle cells in isolation or loose clusters are usually very polymorphic and elongated, along with prominent nucleoli, and solitary, large, and spindled nuclei, as well as a high nuclear–cytoplasmic ratio. The neoplastic giant cells have multilobulated nuclei, abundant eosinophilic cytoplasm, and heavy neutrophilic infiltrate (with occasional ingested white blood cells). Myxoid stroma, numerous mitotic figures, massive necrosis, and vascular invasion (58%) are observed. Immunohistochemically, the epithelial lineage of the spindle and giant cell components in pleomorphic carcinoma is positive for AE1/3, CAM 5.2, CK18, CK7 (76%), TTF1 (59%), and surfactant protein A (39%); the sarcomatoid component is positive for CK7 (63%), TTF1 (43%), and surfactant protein A (6%); both the epithelial and sarcomatoid components are positive for EMA and negative for smooth muscle actin (usually), desmin (usually), HHF-35, calponin, and caldesmon [7,8].

Carcinosarcoma (56.5% of PSC cases) has a mixture of epithelial element (squamous, 45%–70%; adenocarcinomatous, 20%–30%; or large cell, 10%) and mesenchymal/sarcomatous component (rhabdomyosarcoma, osteosarcoma mixed with chondrosarcoma, and osteosarcoma alone). The epithelial component in carcinosarcoma is positive for cytokeratins; the chondrosarcomatous component is positive for S100; and the rhabdomyosarcomatous component is positive for muscle markers.

Pulmonary blastoma (1% of PSC cases) is a biphasic tumor composed of a primitive epithelial component (resembling well-differentiated fetal adenocarcinoma) and a primitive mesenchymal stroma (containing rhabdomyosarcoma, osteosarcoma, or chondrosarcoma). The epithelial component in pulmonary blastoma is positive for cytokeratins, EMA, and carcinoembryonic Antigen (CEA) and variably positive for neuroendocrine markers and specific hormones; the stromal component is positive for vimentin and muscle-specific actin; striated muscle is positive for desmin; and cartilage is positive for S100 [7].

Differential diagnoses for PSC include the subtypes of sarcomatoid carcinomas, malignant mesothelioma, primary or metastatic sarcoma, reactive/inflammatory processes (inflammatory pseudotumour, bronchiolitis obliterans/organizing pneumonia), fibrosarcoma, pleomorphic malignant fibrous histiocytoma, leiomyosarcoma, follicular dendritic cell sarcoma, angiosarcoma, and synovial sarcoma [1].

The stages of PSC are usually determined in accordance with the TNM system of AJCC. A recent survey showed that 8.7%, 15.2%, 19.6%, 32.6%, 15.2%, and 8.7% of patients were diagnosed with stages IA, IB, IIA, IIB, IIIA, and IV PSC diseases, respectively [8].

16.7 Treatment

For localized PSC, surgery (e.g., lobectomy, bilobectomy, pneumonectomy, sleeve lobectomy, or anatomical segmentectomy) remains adequate for treatment and appears to provide adequate local control. For metastatic PSC, treatment is similar to that used for NSCLC. Chemotherapy (e.g., CAV [cyclophosphamide, adriamycin, and vincristine], doxorubicin, ifosfamide, platinum salts, gemcitabine, vinorelbine, paclitaxel, docetaxel, and cisplatin) may have limited benefits for patients with metastatic sarcomatoid carcinoma [3].

16.8 Prognosis

PSC is a poorly differentiated carcinoma with aggressive features and relative refraction to chemotherapy and radiotherapy [8]. Patients with PSC have a median survival time of 12 months after potentially curative surgical resection, and overall survival rates at 1, 2, 3, and 5 years of 45.5%, 35.8%, 28.2%, and 20.1%, respectively [6].

References

1. Franks TJ, Galvin JR. Sarcomatoid carcinoma of the lung: Histologic criteria and common lesions in the differential diagnosis. *Arch Pathol Lab Med.* 2010; 134(1): 49–54.
2. Travis WD, Brambilla E, Burke AP, Marx A, Nicholson AG. *WHO Classification of Tumours of the Lung, Pleura, Thymus and Heart.* International Agency for Research on Cancer, Lyon; 2015.
3. Ouziane I, Boutayeb S, Mrabti H, Lalya I, Rimani M, Errihani H. Sarcomatoid carcinoma of the lung: A model of resistance of chemotherapy. *N Am J Med Sci.* 2014; 6(7): 342–5.
4. Italiano A, Cortot AB, Ilie M, et al. EGFR and K-RAS status of primary sarcomatoid carcinomas of the lung: Implications for anti-EGFR treatment of a rare lung malignancy. *Int J Cancer.* 2009; 125: 2479–82.
5. Romano A, Grassia M, Rossetti AR, et al. Sarcomatoid carcinoma of the lung: A rare case of three small intestinal intussusceptions and literature review. *Int J Surg Case Rep.* 2015; 13: 48–50.
6. Huang SY, Shen SJ, Li XY. Pulmonary sarcomatoid carcinoma: A clinicopathologic study and prognostic analysis of 51 cases. *World J Surg Oncol.* 2013; 11: 252.
7. e-immunohistochemistry.info/website. Sarcomatoid carcinoma of lung. http://e-immunohistochemistry.info/web/Sarcomatoid_carcinoma_of_lung.htm; accessed January 5, 2017.
8. Roesel C, Terjung S, Weinreich G et al. Sarcomatoid carcinoma of the lung: a rare histological subtype of non-small cell lung cancer with a poor prognosis even at earlier tumour stages. *Interact Cardiovasc Thorac Surg.* 2017; 24(3): 407–13.

17
Pulmonary Benign Tumors and Preinvasive Lesions

17.1 Definition

As non-cancerous growths of the lungs that do not spread (metastasize) to other parts of the body and that are usually non-life-threatening, pulmonary benign tumors include hamartoma (representing 55% of benign lung neoplasms), papilloma, adenoma, leiomyoma (1.5%) and lipoma (1.5%).

Pulmonary hamartoma is an overgrowth of cartilage-like cells that is often asymptomatic. Originating from the epithelium in the bronchi, pulmonary papilloma may block the flow of air and consists of squamous cell papilloma (exophytic, inverted), glandular papilloma, and mixed squamous cell and glandular papilloma. Evolved from glandular cells (which produce substances such as mucus, hormones and lubricating fluids) in the lung epithelium, pulmonary adenoma is subdivided into sclerosing pneumocytoma, alveolar adenoma, papillary adenoma, mucinous cystadenoma, and mucous gland adenoma. Leiomyoma is a rare benign lung tumor made up of smooth muscle cells [1,2].

Pulmonary preinvasive lesions are neoplasias that have the potential to progress to malignancy but have not done so yet. They include atypical adenomatous hyperplasia (AAH), adenocarcinoma *in situ* (AIS; nonmucinous/mucinous; formerly bronchioloalveolar carcinoma), squamous cell carcinoma (SCC) *in situ*, and diffuse idiopathic pulmonary neuroendocrine cell hyperplasia (DIPNECH) [1,3,4].

17.2 Biology

Pulmonary hamartoma is made up of "normal" tissues such as cartilage, connective tissue, fat, and muscle but in abnormal amounts. It is found mainly in the peripheral, or outer, portion of the lung's connective tissue (80%) and occasionally inside the bronchial tubes (the airways leading to the lungs) (20%). Previously used interchangeably with pulmonary hamartoma, pulmonary chondroma is now considered by some authors as distinct identity due to its predominant bone component.

Pulmonary papilloma generally occurs as exophytic tumor in the more proximal upper airways and occasionally in the more distal location with an inverted growth pattern. It can be solitary or multiple, with multifocality (multiple papillomas) located in the upper respiratory tract.

Sclerosing pneumocytoma is a benign, asymptomatic, solitary nodule occurring mostly in peripheral lung parenchyma, with an epithelial, type II alveolar cell (pneumocyte) origin. Mucinous cystadenoma is a rare, benign epithelial neoplasm of the lung that often presents as a peripheral asymptomatic cystic nodule [2].

Among pulmonary preinvasive lesions, AAH is a preneoplastic lesion that may progress via AIS (a subtype of the former bronchioloalveolar carcinoma) into peripheral "parenchymal" adenocarcinoma of the lung. Whereas the majority of adenocarcinomas arise in the peripheral airways and lung parenchyma, SCC usually develops in the central airways (first five generations of bronchi) through a stepwise process where the epithelium changes from normal to hyperplasia, metaplasia, mild, moderate, and severe dysplasia and then carcinoma *in situ*. DIPNECH is an exceptionally rare lesion associated with the development of multiple carcinoid tumors [5–7].

17.3 Epidemiology

Pulmonary hamartoma makes up 55% of benign lung tumors, 8% of all lung tumors and 6% of solitary pulmonary nodules. It tends to affect people aged between 50 and 70 years, with a male preponderance.

Solitary respiratory papillomas have a yearly incidence of 3.95 per 100,000, account for <0.5% of all lung tumors and display a male prediction (male: female ratio of 3:1). While squamous papilloma commonly occurs during the fifth decade of life, glandular papilloma predominates in the sixth decade, mixed type papilloma is present from the third to the sixth decade of life.

Sclerosing pneumocytoma has a female predominance with an average age at presentation in the fifth decade (ranging from the second to the eighth decade). Pulmonary mucinous cystadenoma is usually observed in adults of 60–80 years in age.

AAH and AIS are frequently found in cancer-bearing lungs, especially in those with adenocarcinoma. AAH is more common in women, and AIS has a median age at diagnosis of 65 years.

17.4 Pathogenesis

Possible risk factors for pulmonary benign tumors include foreign body, smoking (and passive smoking in females), previous tuberculosis, and human papillomavirus (HPV types 6, 11, 16, 18, 31, 33, and 35). HPV proteins E6 and E7 bind to the tumor suppressor gene p53 and retinoblastoma protein, inducing cellular proliferation and atypia as in the case of lung papilloma. In rare cases, papillomatosis may undergo malignant transformation to become squamous cell carcinoma.

LOH events at 9q and 16p and the mutual exclusion of *KRAS* and *EGFR* gene mutations are found in AAH and lung adenocarcinoma, lending support to the existence of an AAH–AIS–adenocarcinoma continuum. Additionally, AAH displays significantly elevated DNA methylation at *CDKN2A* ex2 and *PTPRN2*, whereas AIS has significant hypermethylation in CpG islands at *2C35*, *EYA4*, *HOXA1*, *HOXA11*, *NEUROD1*, *NEUROD2*, and *TMEFF2* [7].

Apart from tobacco exposure, preinvasive squamous bronchial lesion is linked to LOH at chromosomal loci 3p (3p21, 3p14, 3p22–24, and 3p12) and 9p21 ($p16_{INK4a}$). Other genetic abnormalties associated with pulmonary squqmous dysplasia/carcinoma in situ (CIS) include loss of p16 expression in moderate dysplasia (12%) and CIS (30%); cyclin D1 overexpression in 6% of early hyperplasia/metaplasia, 17% of moderate dysplasia, 46% of moderate dysplasia, and 38% of CIS lesions; aneuploidy in 8% of low grade preinvasive lesions and 33% of high grade preinvasive lesions; Bcl-2 protein expression in 33% of bronchial dysplasias; p53 protein expression in 6.7% of squamous metaplasias; increased telomerase positivity in 70–80% hyperplastic and dysplastic bronchial epithelium, and 95–100% of CIS compared to 20% in normal controls; loss of FHIT (fragile histidine triad) tumor suppressor gene at 3p14.2 in 60% of moderate dysplasia, and 100% in severe dysplasia and CIS [3,7].

17.5 Clinical features

Pulmonary benign tumors (including hamartoma, papilloma and adenoma) located centrally in large airways may induce either endoluminal obstruction or extraluminal compression, leading to persistent cough or wheezing, hemoptysis, progressive dyspnea/difficulty breathing, hoarseness/rattling sounds in the lungs, coughing up blood, chest pain, fever, pneumonia, recurrent infection, pleural effusion, and lung tissue collapse. Some benign tumors localized in the lung parenchyma may be asymptomatic, and are often found by accident during a chest X-ray or CT for unrelated complains.

17.6 Diagnosis

Diagnosis of pulmonary benign tumors and preinvasive lesions involves chest x-ray, CT, fine needle aspiration (FNA), bronchoscopy and histological examination.

Pulmonary benign tumors. Pulmonary hamartoma is usually a slow-growing (typical doubling time >400 days), unencapsulated, tan-white to gray, focal nodule of <4 cm in diameter with glistening, translucent, homogenous, firm to hard cut surface (cartilaginous) and ill defined clefts and connective tissue septa, located in the peripheral lung (>90%) and endobronchial areas (<10%) without pressing against nearby tissue. On chest radiography, it usually presents as a non-specific, coin-like round growth with fluffy wool or popcorn appearance (about 15% of cases). On CT, it has a well-circumscribed solitary mass with 'popcorn' calcification and focal areas of fat (with a Hounsfield measurement of 40–120 HU). On MRI, it gives heterogeneous signal (mainly intermediate signal, foci of high signal representing fat, and low signal regions representing fibrous or calcific material) on T1 weighted images, high signal due to fat and cartilaginous components and low signal regions representing fibrous or calcific material on T2 weighted images, and heterogeneous enhancement on T1 C+ (Gd). Microscopically, pulmonary hamartoma is a disorganized collection of epithelial and other tissue elements (e.g., cartilage, fat, muscle and fibroblasts) with scattered calcification. The presence of only one type of mesenchymal tissue may be indicative of leiomyoma (smooth muscle), lipoma (fat), or chondroma (bone). Immunohistologically, it is positive for vimentin (fibromyxoid cells) but negative for cytokeratins. Molecularly, it contains abnormal karyotypes in mesenchymal component, including aberrations in 12q14–15, 6p21.3, and 14q24 [8,9].

Squamous cell papilloma is the most common pulmonary papilloma in adults. Histologically, the tumor shows exophytic growth although atypical inverted endophytic growth occasionally occurs. The presence of cellular atypia highlights its malignant potential.

Glandular papilloma is a very rare tumor that occurs predominantly in the central tracheobronchial tree. Arising from the mucosal surface, the tumor (0.7–2.5 cm in size) is friable, red to tan in color, and responsible for 40–60% narrowing of the airway lumen. Histologically, the tumor contains ciliated cells or simple columnar cells or mucous cells with a central fibrovascular core. The presence of ciliated cells helps rule out other malignant tumors.

Mixed papilloma occurs predominantly in middle-aged and old smokers (mean age of 58.3 years) with a male preponderance. Typically located centrally, the tumor is composed of squamous and glandular epithelium, with a large seated base and a sharp border. The tumor stains positive for intracellular mucin MUC5AC (which is expressed in tracheobronchial goblet cells), CAM5.2 and CK19 (diffusely positive, indicating its columnar epithelium origin), CEA and CA19-9 (focally positive), CK7, CK5/6 (weak for mucous cells), CK34 βE12 and TTF-1; but negative for CK20.

Sclerosing pneumocytoma is a well circumscribed, nonencapsulated, peripheral mass of about 3 cm in size, gray to tan-yellow in color, with punctuate hemorrhage. Histologically, the tumor is characterized by the presence of two morphologic populations (i.e., cuboidal epithelium resembling type II pneumocytes; uniform round to oval stromal cells with pale eosinophilic cytoplasm) and the formation of four patterns/areas (solid, hemorrhagic, papillary, and sclerotic). The papillary areas consist of the pale round stromal cells lined by the cuboidal epithelium; the solid areas comprise sheets of these round stromal cells; the hemorrhagic areas have hemosiderin accumulation and dilated spaces filled with blood; and the sclerotic areas contain dense collagen. Immunohistochemically, the cuboidal epithelial cells are positive for cytokeratin, EMA, TTF-1, Clara cell 10-kd protein (CC10) and surfactant proteins; the round stromal cells are positive for EMA and TTF-1 as well as vimentin and progesterone receptor, but negative for cytokeratin, C10 and surfactant proteins [2].

Alveolar adenoma is a soft, lobulated nodule containing multiple cysts in the peripheral lung. Histologically, the cysts are lined by a cuboidal and sometimes 'hobnailed' epithelial cells, without papillae.

Mucinous cystadenoma of the lung is a mucinous cystic mass (>90% extracellular mucin) with well-defined margins, Histopathologically, the tumor shows abundant mucin-filling cysts surrounded by thick fibrous walls, which are lined with columnar mucinous epithelial cells. The tumor is positive for TTF-1 and CK7, but negative for CK20.

Pulmonary preinvasive lesions. AAH is the earliest detectable preinvasive lesion that is defined as localized proliferation of cuboidal to columnar epithelial cells in the lining of the alveoli and respiratory bronchioles, leading to mild to moderate degrees of cytologic atypia and focal lesion (called *ground glass nodules*, GGN) of <5 mm in diameter with no solid or part-solid component in the peripheral lung. Low-grade AAH is characterized by a single layer of round to cuboidal cells with low cellular density, small cell size, and minimal variation in nuclear size, shape, and chromaticity, as well as minimal thickening of the alveolar septum. High-grade AAH demonstrates higher cellular density with stratification of the epithelial cells, more significant cytologic atypia, and more extensive fibrosis and thickening of the alveolar septa [3,7].

AIS (nonmucinous or mucinous) is defined as a localized small adenocarcinoma (≤3 cm) involving restricted growth of neoplastic cells along preexisting alveolar structures (lepidic growth), without stromal, vascular, or pleural invasion. Being mostly nonmucinous in which the atypical cells represent Type II alveolar cells or Clara cells, AIS is similar in morphology (increased size, cellular atypia, and cellularity with pronounced cellular crowding and even mild stratification) to high-grade AAH but typically larger than AAH (3 cm vs. 5 mm). Further, AIS has a GGN of higher attenuation than that characteristic of AAH.

SCC *in situ* is a preinvasive lung lesion that shows extension of the disarray to the epithelial surface with malignant cytological features (e.g., hyperchromatism, multiplicity of nucleoli, irregularities of the nuclear membrane, coarsening of chromatin distribution, and discordance of maturation between nucleus and cytoplasm, so-called dyskaryosis), and mitotic figures present through the full thickness.

DIPNECH is preinvasive proliferation of pulmonary neuroendocrine cells that is confined to the respiratory epithelium without penetration through the basement membrane. DIPNECH includes tumorlets (which extend beyond the basement membrane and form lesions <5 mm in greatest dimension) and nodules >5 mm (which are classified as pulmonary neuroendocrine tumors or carcinoids) (see Chapter 14). The association of lung neuroendocrine tumor with multiple nodules in women, together with complaints of chronic cough and wheezing, should raise suspicion of DIPNECH [4].

In general, benign lung nodules differ from malignant lung nodules in growth rate (very slow growth vs. doubling in size every four months or less), calcification or calcium content (high calcium content and smooth/regular shape vs. low calcium content and irregular shape/rough surface).

17.7 Treatment

Surgical resection may be performed on pulmonary benign tumors (including hamartoma) that cause problems or continue to grow, without further treatment, apart from management of any underlying complications related to the nodule such as pneumonia or an obstruction.

Although the standard care for pulmonary preinvasive lesions is surgery, endoscopic minimally invasive, cost-effective techniques (e.g., argon plasma coagulation [APC], cryotherapy, laser, photodynamic therapy [PDT] and intra-luminal irradiation therapy or brachytherapy) are increasingly utilized [10].

17.8 Prognosis

Pulmonary benign tumors including hamartoma generally have an excellent prognosis. AIS and CIS demonstrate five year post-resection survival rates of 100% and 80%–90%, respectively.

References

1. Travis WD, Brambilla E, Burke AP, Marx A, Nicholson AG. *WHO Classification of Tumours of the Lung, Pleura, Thymus and Heart.* International Agency for Research on Cancer, Lyon; 2015.
2. Borczuk AC. Benign tumors and tumorlike conditions of the lung. *Arch Pathol Lab Med.* 2008; 132 (7): 1133–48.
3. Kerr KM. Pulmonary preinvasive neoplasia. *J Clin Pathol.* 2001; 54(4): 257–71.
4. Ishizumi T, McWilliams A, MacAulay C, Gazdar A, Lam S. Natural history of bronchial preinvasive lesions. *Cancer Metastasis Rev.* 2010; 29(1): 5–14.
5. Gazdar AF, Brambilla E. Preneoplasia of lung cancer. *Cancer Biomark.* 2010; 9(1–6): 385–96.
6. Gardiner N, Jogai S, Wallis A. The revised lung adenocarcinoma classification—An imaging guide. *J Thorac Dis.* 2014; 6(Suppl 5): S537–46.
7. Vinayanuwattikun C, Le Calvez-Kelm F, Abedi-Ardekani B, et al. Elucidating genomic characteristics of lung cancer progression from in situ to invasive adenocarcinoma. *Sci Rep.* 2016; 6: 31628.
8. Murphy A, Gorrochategui M, et al. Pulmonary hamartoma. https://radiopaedia.org/articles/pulmonary-hamartoma-1. Accessed January 20, 2017.
9. PathologyOutlines.com website. Hamartoma. http://www.pathology-outlines.com/topic/lungtumorhamartoma.html. Accessed January 21, 2017.
10. Daniels JM, Sutedja TG. Detection and minimally invasive treatment of early squamous lung cancer. *Ther Adv Med Oncol.* 2013; 5(4): 235–48.

18
Small Cell Lung Carcinoma

18.1 Definition

Small cell lung carcinoma (SCLC, or *small cell lung cancer*) is a malignant epithelial tumor characterized by the presence of round, oval, and spindle-shaped small cells with scant cytoplasm, extensive necrosis, high mitotic figures, and neuroendocrine morphology. Compared to other lung neuroendocrine tumors, that is, well-differentiated, Grade I typical carcinoid (TC; representing 2% of primary lung neoplasms), moderately-differentiated, Grade II atypical carcinoids (AC; 0.2%; see Chapter 14, Table 14.1), and poorly differentiated, Grade III large cell neuroendocrine carcinoma (LCNEC; 3%; see Chapter 11), SCLC is a poorly differentiated, Grade III tumor (20%) [1].

Besides pure SCLC, some SCLC may contain areas with non-small cell carcinoma (e.g., adenocarcinoma, squamous cell carcinoma, large cell carcinoma, and occasionally spindle cell or giant cell carcinoma) and are thus referred to as *mixed* or *combined SCLC*. In combined SCLC and large cell carcinoma, a minimum criterion of 10% large cells is required. However, in frank adenocarcinoma or squamous cell carcinoma, no such percentage is mandatory [2].

18.2 Biology

Apparently derived from pluripotent bronchial precursor cells with the ability to differentiate into each of the major histologic types of lung cancer, including neuroendocrine cells, SCLC (formerly *oat cell carcinoma*, *small cell anaplastic carcinoma*, *undifferentiated small cell carcinoma*, *intermediate cell type*, and *mixed small cell/large cell carcinoma*) mostly arises from the central airways, readily infiltrates the bronchial submucosa, and subsequently spreads to the mediastinal lymph nodes, liver, bones, adrenal glands, and brain. It is a very aggressive disease with early mediastinal lymph node involvement [3]. In addition, SCLC may sometimes result from transformation of NSCLC such as pulmonary adenocarcinoma [4].

18.3 Epidemiology

Once accounting for 20%–25% of all newly diagnosed lung malignancies, the incidence of SCLC has declined to 15%–20% in recent years. The current estimated incidence rate of SCLC is 0.2 to 2 cases per 100,000. SCLC typically affects men of 60–80 years (median age of 60 years), with 99% of patients being smokers.

18.4 Pathogenesis

Risk factors for lung cancer including SCLC are tobacco smoking (with 98% of SCLC patients having a smoking history), air pollution, radon gas (a product of uranium decay), uranium, radiation, asbestos, nickel, and chromium exposure. SCLC demonstrates frequent deletions on chromosomes 3p, 4, 5q, 9p, 10q, 13q, and 17p and gains on 3q, 5p, 6p, 8q, 17q, 19, and 20q, in addition to mutations in retinoblastoma and p53 genes. Other notable changes in SCLC include upregulation of the proapoptotic molecule Bcl-2, activation of autocrine loops (bombesin-like peptides, c-kit/stem cell factor), upregulation of telomerase, loss of laminin-5 chains and inhibitors of matrix metalloproteinases, and expression of vascular growth factors. Indeed, insulinoma-associated gene 1 (IA-1) and the human achaete-scute homolog 1 (hASH1) are identified as SCLC markers [5].

The pathogenesis of SCLC may be related to the neuroendocrine activity and autoimmune phenomena. The former increases the production of peptide hormones that contribute to a diversity of paraneoplastic syndromes. The latter is responsible for various neurologic syndromes [2].

18.5 Clinical features

Clinical symptoms of lung cancer range from coughing (smoker's cough), dyspnea (bronchial stenosis, malignant pleural effusion, or pericardial fluid buildup), hemoptysis (small tinges of blood in expectorate), respiratory tract infections (due to a bronchial obstruction), chest pain (due to direct involvement of the chest wall or pleura), loss of appetite, weight loss, to fatigue.

About 10% of SCLC patients manifest with superior vena cava syndrome (dyspnea/stridor, tachypnea, dilated veins on the upper body, edema of the face and neck/arms, chest pain, difficulty in swallowing, vocal cord paralysis); 10% develop abnormal adrenocorticotropic hormone (ACTH)-like activity; and 15% produce antidiuretic hormone (ADH) (inappropriate ADH syndrome, Schwartz–Bartter syndrome) leading to edema, and clumsy, tired, and weak feelings [2].

18.6 Diagnosis

Diagnosis of lung cancer involves medical history review (particularly smoking habits/patterns, work-related exposures, and weight changes), clinical examination (e.g., EKG, spirometry, diffusion capacity, walking test, and cycling test), imaging (CT scan, PET-CT, and MRI of thorax and upper abdomen), bronchoscopy with biopsy (for suspected centrally located tumor on a CT or clinical sign of lung cancer), cytological and histological examination of fine-needle aspirate or biopsy specimen.

Macroscopically, SCLC is a white-tan, soft, friable, perihilar mass with extensive necrosis and frequent nodal involvement. On imaging, SCLC appears as hilar or perihilar mass often with mediastinal lymphadenopathy and lobar collapse. On CT, SCLC shows mediastinal nodal involvement and superior vena cava obstruction.

Microscopic examination of H&E stained SCLC aspiration biopsies and brushings reveals sheets, ribbons, clusters, rosettes, or peripheral palisading of small to medium-sized (2–4 times the size of neutrophils), round to fusiform cells with scant cytoplasm, finely granular and uniformly distributed chromatin ("salt and pepper" quality), hyperchromatic, inconspicuous nucleoli, nuclear molding, smudging, average 80 mitotic figures per 2-mm^2 area; Azzopardi effect (basophilic nuclear chromatin around blood vessels, especially in necrotic areas), indistinct cell borders; scanty, vascular, delicate stroma; common necrosis and apoptotic debris; and possibility of larger cells with similar morphology, squamous cell carcinoma, or adenocarcinoma (as in combined SCLC). Immunohistochemically, the tumor is positive for pancytokeratin (100%, dot-like pattern), thyroid transcription factor-1 (TTF1, 89%), neuron-specific enolase (77%), CD117 (50%–75%), chromogranin (58%), synaptophysin (57%), calretinin (49%), thrombomodulin (27%), keratin5 (27%), CD57/Leu7 (variable), gastrin-releasing peptide, CD56 (also known as nucleosomal histone kinase 1 or neural-cell adhesion molecule), and bcl-2 (variable), but negative for CD3, CD20, CD45, CD99/MIC2, pancreatic polypeptide, vimentin, mesothelin, and p63 [6].

Differential diagnoses encompass other neuroendocrine tumors (i.e., typical carcinoid tumor, ≥0.5 cm in size, with <2 mitoses per 2 mm^2, no necrosis; atypical carcinoid tumor, 2–10 mitotic figures per 2 mm^2, focal necrosis, less nuclear atypia, more intense neuroendocrine staining; LCNEC, mean 70 mitoses per 2 mm^2, large zones of necrosis), other small, round, blue cell tumors (such as primitive neuroectodermal tumor, which is keratin negative and CD99 positive), primary or metastatic non-small cell carcinomas (e.g., lymphoma and melanoma, which are positive for keratin expression and negative for squamous markers such as p63), and lymphoid infiltrates (due to chronic inflammation or small lymphocytic lymphoma) [2].

The stage of SCLC is determined by using the TNM system, which incorporates inputs from bronchoscopy, chest X-ray, chest CT scan, and upper abdominal CT scan or ultrasonography, plus a bone marrow examination and/or a bone scintigram (or PET scan) as well as MRI if bone metastases or central nervous system metastases are suspected. Within the TNM system, *T* (*tumor*) describes the size and extent of the local tumor; *N* (*node*) indicates whether there is metastasis to lymph nodes; and *M* (*metastasis*) indicates the absence or presence of distant metastasis. The TNM system provides a guide for the choice of treatment. In case distant metastases (M1) are present, the T and N stages are of minor importance [7,8].

SCLC with metastasis is often separated into limited or extensive disease. Limited disease (SCLC-LD) is restricted to one hemithorax with regional lymph node metastases (including hiliar ipsilateral and contralateral; mediastinal ipsilateral and contralateral; supraclavicular ipsilateral and contralateral; and ipsilateral pleural effusion) and is equivalent to Stages I–III of the TNM system, with the tumor masses found in a thoracic radiation field. Extensive disease (SCLC-ED) is equivalent to Stage IV in the TNM system and includes spread of malignant cells outside the thoracic radiation field or pleural effusion (e.g., the contralateral lung [10%], skin or distant lymph nodes [10%], brain [10%], liver [25%], adrenals [15%], bone marrow [20%], retroperitoneal lymph nodes [5%], pancreas [5%], and rarely osteolytic bone) [2].

18.7 Treatment

Treatment for lung cancer includes surgery, radiation therapy, or chemotherapy (typically a combination of platinum and etoposide), either alone or in different combinations. Use of steroids, sedative and pain medication, drainage of pleural fluid, and transfusions may be considered if necessary [8].

In limited disease (SCLC-LD), where the cancer is confined to one hemithorax, treatment with combination chemotherapy and thoracic radiation therapy has cure rates of 15%–25%. For small peripheral tumors, surgery is considered, followed by chemotherapy (so called chemoradiation therapy). Other treatment options for SCLC-LD are combination chemotherapy alone or prophylactic cranial irradiation.

In extensive disease (SCLC-ED), which is incurable, treatment options consist of combination chemotherapy (including palliative therapy for compression of airways, superior vena cava syndrome [compression of central venous blood vessels], and spinal cord compression syndrome [tumors that can damage the spinal cord], radiation therapy, and prophylactic cranial irradiation.

For recurrent SCLC, treatment options include chemotherapy and palliative therapy.

Given the universal bi-allelic inactivation of TP53 and RB1 as well as the presence of other genetic alterations in SCLC, use of checkpoint blockers (e.g., ipilimumab/nivolumab, pembrolizumab, or tremelimumab/durvalumab) offers potentially promising, immunological alternatives for combating this aggressive, rapid-growing, early-recurring, and frequently metastatic malignancy [9].

18.8 Prognosis

Patients with SCLC-LD have a complete response rate of 80% to chemotherapy and chest radiation and survive for 17 months. About 15% of SCLC-LD patients are alive at 5 years.

Patients with SCLC-ED have a complete response rate of >20% to chemotherapy and a median survival of >7 months. Only 2% of SCLC-ED patients are alive at 5 years.

Poor prognostic factors include relapsed disease; weight loss; hyponatremia; elevated serum LDH, alkaline phosphatase, albumin, hemoglobin, white blood count; extensive stage of disease; low plasma albumin and sodium levels; single-nucleotide polymorphisms within the promoter region of YAP1 on chromosome 11q22; and phosphorylated Bcl-2 and Mcl-1 [10]

References

1. Travis WD, Brambilla E, Nicholson AG, et al. The 2015 World Health Organization classification of lung tumors: Impact of genetic, clinical and radiologic advances since the 2004 classification. *J Thorac Oncol.* 2015; 10(9): 1243–60.
2. Travis WD. Update on small cell carcinoma and its differentiation from squamous cell carcinoma and other non-small cell carcinomas. *Mod Pathol.* 2012; 25 Suppl 1: S18–30.

3. Semenova EA, Nagel R, Berns A. Origins, genetic landscape, and emerging therapies of small cell lung cancer. *Genes Dev.* 2015; 29 (14): 1447–62.

4. Dorantes-Heredia R, Ruiz-Morales JM, Cano-García F. Histopathological transformation to small-cell lung carcinoma in non-small cell lung carcinoma tumors. *Transl Lung Cancer Res.* 2016; 5(4): 401–12.

5. Mlika M, Laabidi S, Afrit M, Boussen H, El Mezni F. Genomic classification of lung cancer: Toward a personalized treatment. *Tunis Med.* 2015; 93(6): 339–44.

6. PathologyOutlines.com website. *Small cell carcinoma.* http://www.pathologyoutlines.com/topic/lungtumorsmallcell.html;   accessed January 8, 2017.

7. Liam CK, Andarini S, Lee P, Ho JC, Chau NQ, Tscheikuna J. Lung cancer staging now and in the future. *Respirology.* 2015; 20(4): 526–34.

8. PDQ Adult Treatment Editorial Board. *Small Cell Lung Cancer Treatment (PDQ®): Health Professional Version.* PDQ Cancer Information Summaries [Internet]. Bethesda (MD): National Cancer Institute (US); 2002-. Published online January 20, 2017.

9. Patel SH, Rimner A, Cohen RB. Combining immunotherapy and radiation therapy for small cell lung cancer and thymic tumors. *Transl Lung Cancer Res.* 2017; 6(2): 186–95.

10. Chowdry RP, Sica GL, Kim S. Phosphorylated Bcl-2 and Mcl-1 as prognostic markers in small cell lung cancer. *Oncotarget.* 2016 Feb 18. doi: 10.18632/oncotarget.7485.

SECTION III
Digestive System

19
Anal Cancer

Zainul A. Kapacee and Shabbir Susnerwala

19.1 Definition

Cancers of the anal canal occur in the distal portion of the lower gastrointestinal tract, immediately inferior to the rectum. The anal canal is 30–40 mm in size beginning at the superior aspect of the pelvic diaphragm and terminates at the anus (Figure 19.1).

The well-established tumor–node–metastasis (TNM) classification system for solid tumors is widely used to define and stage anal cancers, as shown in Table 19.1.

19.2 Biology

The anal canal consists of three histological types of mucosa, which from proximal to distal are glandular, transitional, and squamous epithelium (Figure 19.1). Glandular epithelium gives rise to adenocarcinomas of the anal canal, which are rare, accounting for <5% of anal cancers. Histologically, these cancers resemble colorectal adenocarcinomas and are thus managed using treatment regimes similar to those for colorectal cancers.

The transitional zone, anatomically located at the dentate line, typically form basaloid cell or squamous cell tumors. Below the dentate line, the anal canal is lined by squamous epithelium, which merges with the perianal skin distally. Tumors from this region are mostly squamous cell carcinomas (SCC), which account for >95% of all anal canal cancers. The common term *anal cancer* refers to this histological subtype.

Of note, tumors arising from the hair-bearing perianal skin are referred to as *anal margin tumors*. Typically these tumors form well-differentiated SCC but clinically can be very difficult to distinguish from tumors arising from the anus/anal canal or a discrete skin lesion.

19.3 Epidemiology

Anal cancer is a relatively uncommon malignancy, with incidence of 1.8 per 100,000 per year. Anal cancer is more common in females, typically

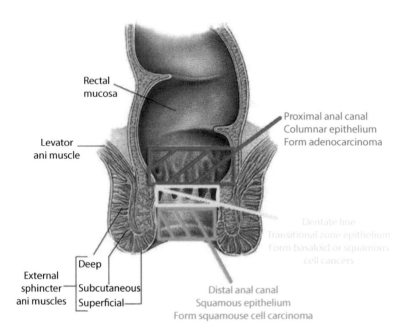

Rectal
mucosa

Proximal anal canal
Columnar epithelium
Form adenocarcinoma

Levator
ani muscle

Dentate line
Transitional zone epithelium
Form basaloid or squamous
cell cancers

Deep

External
sphincter — Subcutaneous
ani muscles — Superficial

Distal anal canal
Squamous epithelium
Form squamouse cell carcinoma

Figure 19.1   The anal canal. The highlighted areas demonstrate the histologically distinct thirds of the anal canal.

Table 19.1   TNM Classification of Anal Canal Cancers

Tumor	Node	Metastasis
Tx: Tumor cannot be assessed	Nx: Nodes cannot be assessed	M0: No distant metastases
T0: No evidence of tumor	N0: No nodal involvement	M1: Distant metastases
Tis: Anal intraepithelial neoplasia (AIN): Dysplastic changes of confined epithelial cells	N1: Perirectal node involvement	
T1: Tumor <2 cm in size	N2: Unilateral internal iliac and/or inguinal node involvement	
T2: Tumor >2 cm but <5 cm in size	N3: Bilateral internal iliac and or inguinal node involvement	
T3: Tumor >5 cm in size		
T4: Tumor of any size but invading adjacent tissues or viscera		

Note: TNM, tumor–node–metastasis.

accounting for 60%–65% of all anal cancers. The prevalence of anal cancer increases with age with >50% of all diagnoses in patients 60 years and older. With changes in sexual practice, namely anal sex, and spread of human papilloma virus (HPV) infections, an increasing number of cases are being diagnosed in younger patients [1,2].

19.4 Pathogenesis

HPV infection is a well-known risk factor for SCC particularly of the cervix and oropharynx, with 60%–90% of all anal SCCs associated with HPV. HPV Types 6 and 11 are the most documented low-risk viruses and typically cause benign lesions such as anogenital warts. HPV Types 16 and 18 are high-risk variants and strongly associated with anal cancers due to their pro-oncogenic abilities. Patients infected with HIV leading to AIDS or encumbered with other immunocompromising conditions are at greater risk of developing cancers [2].

19.5 Clinical features

The two most common symptoms of anal cancer are perianal pain and perirectal bleeding, which are frequently associated with anal fissures, hemorrhoids, and other benign conditions, complicating the diagnosis of early malignant disease within the anal canal.

Another symptom of anal cancers is a change in bowel habit. Where there is involvement of the anal sphincters (internal and/or external), patients may present with fecal incontinence. Tenesmus, a feeling of incomplete evacuation of stools, can also occur where there is a partially obstructing tumor in the anal canal.

Other general symptoms that patients may present with include fatigue and shortness of breath due to anemia, weight loss, and rarely symptoms of venous thromboembolism (calf pain, pleuritic chest pain, shortness of breath). Finally, a number of patients may be asymptomatic and are only detected incidentally.

19.6 Diagnosis

The gold standard investigation for diagnosing anal cancers is examination under anesthesia of the anal canal and rectum and tissue biopsy for histological confirmation of malignancy.

MRI of the pelvis is a vital step in the diagnostic process. MRI scanning provides detailed soft tissue images required to anatomically define the anal tumor size and involvement with surrounding structures, as well as to identify the presence of nodal disease in the mesorectum and inguinal region. The detailed images obtained are essential in the planning of radiotherapy. MRI scans are also used for posttreatment follow-up, to compare with previous scans, to evaluate the therapy response (Figure 19.2).

CT scanning of the thorax, abdomen, and pelvis is performed for accurate clinical staging of disease. Where available, some centers also offer

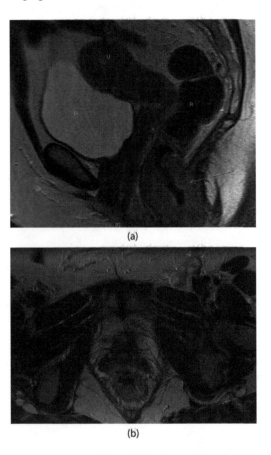

(a)

(b)

Figure 19.2   MRI scan images of the anal canal for diagnosis and treatment planning. (a) shows a sagittal plane of the pelvis; B, U, and R indicate bladder, uterus, and rectum, respectively. The arrow indicates the tumor within the anal canal. (b) shows a transverse section of the pelvis. The arrowheads at the top left of the scan highlight metastatic necrotic inguinal lymph nodes.

positron-emission topography (PET) scanning. In this test, a radioactively labelled metabolite such as glucose is given intravenously and is readily taken up by cells with high metabolic activity such as cancer cells. PET scanning is mapped to a CT scan to identify areas of high tumor activity. Integrated PET-CT in anal cancers are particularly useful for the detection of metastatic inguinal node involvement. However, this imaging modality is expensive and hence, at present, its use is limited.

As part of the workup, all patients would require a full medical history and examination, routine blood tests including full blood count, renal and hepatic function and electrolytes, electrocardiogram, and gynecological examination in women prior to commencement of treatment. At present there are no sensitive or specific serum tumor markers for anal SCCs to aid diagnosis.

19.7 Treatment

Up until the 1970s the mainstay of treatment was localized excision of early disease or more commonly abdominoperineal resection (APR) surgery of the anal cancer and anus for more advanced disease. However, incidence of disease recurrence and spread occurred in over 60% of patients, conferring poor survival outcomes. In addition, patients required a permanent colostomy bag after APR surgery, with resultant impacts on quality of life. Nowadays APR surgery is used strictly as a salvage treatment when current chemotherapy and radiotherapy regimes have failed to achieve local disease control or where there is disease recurrence.

The treatment of choice at present is concurrent chemotherapy and radiotherapy, with most treatment regimes given over an average 38-day period [3,4]. The antimetabolite chemotherapy agent 5-fluorouracil (5-FU) is delivered as a continuous infusion over Days 1–4 and 29–32 of therapy. Some centers also offer capecitabine, which is taken as an oral tablet instead of 5-FU, making it more convenient to administer to patients. A second chemotherapy agent is mitomycin C, given as a single bolus dose on Day 1 of treatment, being the agent of choice. Cisplatin is equally effective as an alternative to mitomycin, given as two infusions over the course of treatment [4].

Use of radiotherapy began following pioneering work in 1974 by Nigro et al., who found radiotherapy to be equally effective to surgery, the treatment of choice at the time, in treating anal SCCs. Radiotherapy is the treatment using X-rays, delivering high-energy photons by a linear accelerator to the tumor site. Ionized atoms are generated within the irradiated cells, which subsequently cause molecular disruption and DNA damage. Cancer cells are rapidly dividing with poor DNA repair mechanisms, thus making them more vulnerable to radiotherapy.

Radiotherapy is effective in inducing tumor cell death; however, the radiation delivered also causes damage to the normal cells of tissues and viscera surrounding the tumor. In anal cancers, the bladder and bowel receive radiation leading to post-therapy toxicity and side effects. Over the years, there have been numerous advances in radiotherapy delivery, aiming to reduce the dose delivered to the normal tissues. When first introduced, two-dimensional parallel pair radiation fields in anterior–posterior fields were used. This method irradiated the anal cancer tumor as well as the entire pelvic region and viscera at an equal radiation dose, leaving patients with long-term skin changes and bladder and bowel problems. Three-dimensional radiotherapy was subsequently developed, which allowed more planes in which radiation could be delivered to the pelvic field. This method enabled delivery of greater radiation doses to the tumor site and reduced dose delivery to the nontarget viscera.

Currently, intensity modulated radiotherapy (IMRT) is the radiation technique widely adopted to treat anal cancers. This method allows individual beams of radiation to be modified by dose and angulation, which means larger doses of radiation can be delivered to the tumor and involved lymph nodes with far less dose delivered to the normal surrounding tissues. At present there are Phase II and Phase III randomized control trials underway in the UK to investigate the optimal radiation dose using IMRT with dose reduction for early stage tumors and increased dose for locally advanced disease [3].

19.8 Prognosis

Five-year survival rates for anal cancer in the 1950s when APR surgery was the mainstay of treatment ranged from 40% to 60%. Current studies now report 5-year overall survival rates in excess of 80%, particularly for patients with early stage disease.

Advanced tumor stage, as in T3 and T4 tumors, and presence of nodal disease have been shown to confer poorer prognostic outcome. The anal canal circumference involved, measured during clinical exam at initial presentation, has been found to be of prognostic significance, with the greater the circumference of anal canal involved the greater risk of disease recurrence and poorer survival.

Low pretreatment hemoglobin levels have also been shown to confer greater risk of recurrence and poor survival. The hemoglobin level can give an indication towards the patient's general heath; anemic patients may suffer with other comorbidities affecting their prognosis. Furthermore, radiotherapy requires adequate oxygenation of tumor tissues to generate damage by oxygen free radicals within cancer cells, and this may be hampered in anemic patients. Male gender, advancing age, and black ethnicity are reported as poorer prognostic factors.

The focus for the future in anal cancer therapy is to improve the local control of disease using modern radiotherapy techniques while reducing the associated morbidity from radiotherapy to normal tissues.

References

1. Anal cancer statistics. Cancer Research, UK. http://www.cancerresearchuk.org/health-professional/cancer-statistics/statistics-by-cancer-type/anal-cancer; accessed October 1, 2016.
2. Yim, E. K. and J. S. Park. The role of HPV E6 and E7 oncoproteins in HPV-associated cervical carcinogenesis. *Cancer Res Treat.* 2005; 37(6): 319–24.
3. Glynne-Jones, R., et al. Squamous-cell carcinoma of the anus: Progress in radiotherapy treatment. *Nat Rev Clin Oncol.* 2016; 13(7): 447–59.
4. Kapacee, Z. A., et al. Chemoradiotherapy for squamous cell anal carcinoma: A review of prognostic factors. *Colorectal Dis.* 2016; 18(11): 1080–6.

20
Appendiceal Cancer

20.1 Definition

Both primary and secondary tumors are known to affect the appendix. The 2010 *WHO Classification of Tumours of the Digestive System* includes epithelial tumors, mesenchymal tumors, and lymphomas among primary appendiceal neoplasms [1].

Appendiceal epithelial tumors comprise adenoma (tubular, villous, tubulovillous, serrated), carcinoma (adenocarcinoma, mucinous adenocarcinoma, signet-ring cell adenocarcinoma, undifferentiated carcinoma), carcinoid (well-differentiated endocrine neoplasm; enterochromaffin (EC) cell, serotonin-producing neoplasm; L-cell, glucagon-like peptide and PP/PYY producing tumor), tubular carcinoid, goblet cell carcinoid (mucinous carcinoid), and mixed carcinoid–adenocarcinoma. Appendiceal mesenchymal tumors consist of neuroma, lipoma, leiomyoma, gastrointestinal stromal tumor, leiomyosarcoma, and Kaposi sarcoma.

It is noteworthy that appendiceal adenoma is usually confined to the mucosa without mucin or cells penetrating the muscularis mucosa or evidence of perforation and thus is regarded as benign and noncancerous. Appendiceal carcinoma and carcinoid are capable of cellular invasion beyond the muscularis mucosa and thus are considered malignant and cancerous.

The most common primary appendiceal cancer is carcinoid, which accounts for about 65% of all appendiceal malignancies. The remaining cases of appendiceal tumors are represented by mucinous adenocarcinoma (MAC, or *mucinous cystadenocarcinoma*, 20%), adenocarcinoma (10%), and goblet cell carcinoid (GCC, also referred to as *adenocarcinoid, goblet cell adenocarcinoma, mucinous carcinoid, microglandular goblet cell carcinoma*, or *crypt cell carcinoma*; 5%)

20.2 Biology

The human appendix is a narrow pouch of tissue extending from the cecum located in the lower right-hand abdominal area. It measures about 10 cm in length and 0.6 cm in diameter. Structurally, the appendix is composed of four layers: mucosa, submucosa (with masses of lymphoid tissue), muscularis externa, and serosa. The mucosa of the appendix is studded with pits (called the *crypts of Lieberkuhn*), which are lined by a single layer of columnar cells. The crypts also contain mucus-producing, large clearer cells (or *goblet cells*). At the base of the crypts are intestinal stem cells, with the capacity to differentiate into columnar cells (entrocytes), goblet cells, neuroendocrine cells, and Paneth cells. The appendix differs from the colon in having a very large number of lymphoid aggregates in the submucosa. These aggregates produce secretory IgA antibodies to assist in detoxification within the gut. The mucus and antibodies produced by the appendix are subsequently pushed into the cecum by peristalsis.

Although adenoma is often limited to the mucosa of the appendix and does not produce murin, some neoplasms show adenomatous growths with mucin dissection beyond the muscularis mucosa (so-called broad front invasion), mural perforation, or even peritoneal dissemination (as in the case of pseudomyxoma peritonei, PMP). The 2010 WHO classification recognizes three categories of mucinous neoplasms: mucinous adenoma (restricted to mucosa, no extra-appendiceal mucin, not associated with PMP, nonrecurrent), low-grade appendiceal mucinous neoplasm (LAMN; non-invasive glands beyond appendix, extra-appendiceal mucin, associated with low-grade PMP, recurrent), and appendiceal adenocarcinoma (invasive glands beyond appendix, extra-appendiceal mucin, associated with high-grade PMP, 10-year survival <10%) [1,2].

Most primary adenocarcinomas of the appendix (e.g., adenocarcinoma and MAC) evolve from adenomatous polyp or serrated adenoma. MAC may induce a diffuse thickening of the appendix and occlude the narrow lumen, leading to mucin buildup and dilatation of the appendix (so-called mucocele). Simple mucocele is an uncommon retention cyst of <2 cm. Perforation of mucocele spills cancerous cells out into the peritoneum, creating the condition of PMP (which contains mucin pools with low or high cellularity, bland cytology or moderate/severe cytologic atypia, and nonstratified simple cuboidal epithelium or cribriform/signet ring morphology with desmoplastic stroma). Compared to adenocarcinoma found in other parts of the body, appendiceal adenocarcinoma often has a less aggressive natural history.

Carcinoids arise from neuroendocrine cells located in the small intestine (44.7%), rectum (19.6%), appendix (16.7%), colon (10.6%), and stomach (7.2%). Noted by their serotonin production, carcinoids and tumors with neuroendocrine elements are collectively referred to as *neuroendocrine tumors* (NET). Thus, appendiceal carcinoid, tubular carcinoid, GCC, and mixed carcinoid–adenocarcinoma are also NET. Nonetheless, GCC appears to be a distinct type of NET (in both clinical and pathologic terms) [3].

20.3 Epidemiology

Appendiceal tumors account for 0.4%–1% of all gastrointestinal malignancies and have annual incidence of 1.2 cases per 100,000. Appendiceal NET (e.g., carcinoid) are found in 0.3%–0.9% of patients undergoing appendectomy, with a mean age of 38 years at diagnosis. Appendiceal adenocarcinoma consists of three subtypes: mucinous (55%), colonic type (34%), and adenocarcinoid (11%, or GCC), which together represent 0.5% of all gastrointestinal cancers. The mean age at diagnosis is 58 years, which is about 20 years later than the peak age for appendiceal NET.

20.4 Pathogenesis

Appendiceal cancer is linked to mutations in the *APC, ATM, KRAS, IDH1, NRAS, PIK3CA, SMAD4,* and *TP53* genes. Specifically, GCC demonstrates loss in chromosomes 11q, 16q, and 18q, in addition to strong carcinoembryonic antigen, caudal-type homeobox transcription factor 2, cytokeratin 7 (CK7), CK20 expression (an epithelial element not present in classic carcinoids), and transcription factor Math-1 and HD5 expression. MAC and LAMN contain mutations in the *KRAS, GNAS,* and *TP53* genes; losses of SMAD4 protein expression (and loss of heterozygosity at chromosome 18q); and p53 overexpression. In addition to alterations in *KRAS, GNAS,* and *TP53 genes,* mucocele harbors mutations in *JAK3* (Janus kinase 3), AKT1 (v-akt murine thymoma viral oncogene homolog 1), A*PC,* MET (met protooncogene), *PIK3CA,* RB1 (retinoblastoma 1), and STK11 (serine/threonine kinase 11) genes; whereas PMP has mutation in the RB1 gene [3].

20.5 Clinical features

Appendiceal tumors often cause appendicitis or rupture of appendicitis, with acute pain and tenderness in the right lower quadrant as typical symptoms. Other clinical manifestations include nausea, vomiting, fever, change in bowel habits, intussusception, tachycardia, hypotension, ascites, and weight loss. Applying pressure to the area often leads to pain that may sharpen after releasing the pressure (so-called rebound tenderness). In MAC, continuing accumulation of mucin in the peritoneum distends the abdomen, leading to "jelly belly" (or PMP).

20.6 Diagnosis

Patients suspected of appendicitis are often examined by CT; and those suspected of carcinoid syndrome are tested for urinary excretion of

5-HIAA and serum chromogranin A levels. Complete colonoscopy may be used to evaluate synchronous colorectal lesions. Final diagnosis of appendiceal cancer is made upon histological assessment of appendix after appendectomy.

Appendiceal carcinoid is a gray or yellow, well-demarcated, firm, intramural nodule of <1 cm, typically located in the tip of the appendix and associated with narrowed or obliterated lumen. Histologically, the tumor is well differentiated and demonstrates characteristic cytologic and architectural features. These include round, regular nuclei and stippled or salt and pepper chromatin; moderate to abundant cytoplasm; various growth patterns, such as insular (round nests of cells, palisading), trabecular (rows and strands of cells, "stacked" with long axis perpendicular to the long axis of the row), tubular (lined by a single layer of uniform cells, occasional rosettes); involvement of only base of crypts with largely intact mucosa. The tumor is divided into three histologic subtypes: classic, tubular, and goblet cell type. Classic carcinoid (or EC cell serotonin-producing carcinoid) is the most common subtype, characterized by predominantly insular growth pattern; production of serotonin and substance P; positivity for synaptophysin, chromogranin A, argentaffin and argyrophil; and frequent invasion. Tubular carcinoid (or L-cell glucagon-like peptide and pancreatic polypeptide [PP/PYY] producing carcinoid) is an uncommon subtype characterized by predominantly tubular and/or trabecular growth pattern; production of GLP-1, GLP-2, glicentin, oxyntomodulin, and PP/PYY; positivity for synaptophysin, chromogranin B, and argyrophil but negativity for chromogranin A and argentaffin; no invasion, <3 mm [3].

GCC often induces appendiceal wall thickening without the presence of gross tumor. The tumor usually spares mucosa and infiltrates muscularis propria and periappendiceal fat, and demonstrates divergent differentiation of normal crypt stem cells (columnar, endocrine, goblet cell, Paneth cell lineages) between typical carcinoma and adenocarcinoma. The tumor shows solid tumor cell clusters, crypt-like structures or tubules of mucus-secreting cells distended with mucin and also with eosinophilic cytoplasm; bland and monomorphic nuclei; pools of extracellular mucin; scattered Paneth cells in tumors with crypt-like structures; and extensive perineural invasion. Immunohistochemically, the endocrine cell component of GCC is positive for chromogranin A, serotonin, enteroglucagon, somatostatin, and/or PP; the goblet cells express CEA. Molecularly, carcinoid tumor is often 8q-, less frequently 11q-, and usually diploid by flow cytometry [3].

Appendiceal adenocarcinoma is a cystic (cystadenocarcinoma) or noncystic mass that may bury the appendix. Histologically, the tumor consists of intestinal, mucinous, or signet-ring cell types, with mucinous cases typically presenting with PMP. MAC of the appendix often shows the pathognomonic volcano sign (an erythematous soft cecal mass with central crater from which mucin is discharged), and increased incidence of synchronous or metachronous colonic polyps and masses as revealed by endoscopy. On ultrasound, MAC usually presents as an encapsulated, elongated or ovoid cystic lesion attached to the cecum with an internal onion-skin appearance (reflecting lamellated mucin). On T2 weighted images, MAC demonstrates hyperintense tubular distention of the appendix with bright intraluminal and peri-appendiceal mucin. Histologically, the tumor contains malignant cells (destructive invasion of the appendiceal wall, extensive full-thickness nuclear stratification, vesicular nuclei, marked nuclear membrane abnormalities, prominent nucleoli, frequent mitotic figures, complex papillary fronds, cribriform glandular spaces) in wall with desmoplastic stromal response (high-grade MAC). The tumor may be referred to as LAMN in cases of well-differentiated mucinous neoplasm without evidence of frank invasion (presence of undulating or broad front) or absence of tumor cells in mucin deposits. MAC stains positive for CK18, CK20, CEA (which is of diagnostic and prognostic value), CDX2, and MUC2 and may be 18q- at molecular level.

20.7 Treatment

Surgical excision offers an effective treatment for appendiceal tumors, both benign and malignant.

Treatment for carcinoid of <2 cm confined to the appendix is a simple appendectomy without follow-up. Treatment for carcinoid of >2 cm, or with positive surgical margin or extremely pronounced mesoappendiceal spread, is an appendectomy with right hemicolectomy and cytoreductive surgery as well as a 3-month postoperative follow-up (including history and physical, CT, and tests for 5-HIAA and chromogranin A). For patients with metastatic carcinoid, somatostatin analogs (somatostatin and octreotide) may be utilized to relieve the symptoms of carcinoid syndrome.

Treatment of appendiceal adenocarcinoma is appendectomy if well differentiated and superficial, or right hemicolectomy. The colonic and goblet cell subtypes are invasive, and approximately half the patients present with nodal metastasis. GCC may be similarly treated with appendectomy, or right hemicolectomy if predominant carcinomatous growth pattern, significant mitotic activity, or tumor beyond appendiceal wall occurs [4].

20.8 Prognosis

Carcinoid tumors confined to the appendix have a 5-year survival rate of 94% compared to 34% for those with distant metastasis. GCC has an intermediate prognosis between classical carcinoid and adenocarcinoma, with a 5-year survival rate of 55% (Stage I [100%], Stage II [76%], Stage III [22%], and Stage IV [14%]). Poor prognostic factors for GCC include solid pattern or carcinomatous pattern in >50% of tumor, perineural invasion, lymphatic invasion, spread beyond appendix, serosal involvement at presentation, and incomplete excision at appendix base. MAC has a relatively favorable prognosis.

References

1. Bosman FT, Carneiro F, Hruban RH, Theise ND. *WHO Classification of Tumours of the Digestive System*. 4th edition. World Health Organization; International Agency for Research on Cancer. Lyon: IARC Press; 2010.
2. Misdraji J. Mucinous epithelial neoplasms of the appendix and pseudomyxoma peritonei. *Mod Pathol*. 2015; 28 Suppl 1: S67–79.
3. Shenoy S. Goblet cell carcinoids of the appendix: Tumor biology, mutations and management strategies. *World J Gastrointest Surg*. 2016; 8(10): 660–9.
4. Kelly KJ. Management of appendix cancer. *Clin Colon Rectal Surg*. 2015; 28(4): 247–55.

21
Bile Duct Cancer

21.1 Definition

Tumors affecting the bile ducts range from adenoma, carcinoma *in situ*, adenocarcinoma, mucinous adenocarcinoma, clear cell adenocarcinoma, signet-ring cell carcinoma, adenosquamous carcinoma, squamous cell carcinoma, small cell/oat cell carcinoma, to neuroendocrine tumors and lymphoma. Out of these, the most common and important is an epithelial adenocarcinoma called *cholangiocarcinoma* (CC), which accounts for >95% of all tumors involving bile ducts. For this reason, CC is often used interchangeably with *bile duct cancer* in both the literature and colloquial language.

Depending on the location and histological characteristics, CC is subdivided into intrahepatic, perihilar, and distal subtypes. Intrahepatic CC (ICC) is a well-differentiated adenocarcinoma that demonstrates various growth patterns and represents about 10% of all bile duct neoplasms. Perihilar CC (PCC, also known as Klatskin tumor) is a moderately differentiated adenocarcinoma (mostly of mucinous type), comprising about 50% of all bile duct neoplasms. Distal CC (DCC) is a poorly differentiated adenocarcinoma, accounting for about 40% of all bile duct neoplasms.

Collectively, PCC and DCC are referred to as *extrahepatic bile duct cancer* (ECC), which typically forms glands or secretes significant amounts of mucins, and which often shows sclerosing (>70%), nodular (20%), and papillary (5%–10%) growth patterns. The sclerosing and nodular patterns involve the rich lymphatic plexus around the bile ducts in early disease, whereas the papillary pattern displays endobiliary growth and a more favorable prognosis [1].

21.2 Biology

The bile ducts are thin tubes of 10–12.5 cm in length that connect the liver, gallbladder, and small intestine. The bile ducts are divided into two sections: intrahepatic and extrahepatic.

The intrahepatic bile ducts are small ducts located within the liver. These small ducts merge to form the left and right hepatic ducts, which exit the liver at the hilum and then join together to form the common hepatic duct. The extrahepatic bile ducts include part of the right and left hepatic ducts outside the liver, the common hepatic duct, and the common bile duct and may be further divided into the perihilar bile duct (or the proximal extrahepatic bile duct, which starts from the perihilar area and ends at the point where the cystic duct joins the common hepatic duct) and the distal extrahepatic bile duct (located between the junction where the cystic duct joins the common hepatic duct and the ampulla of Vater, which is a channel formed by the common bile duct joining the pancreatic duct). The cystic duct connects the common bile duct with the gallbladder.

The main function of the bile ducts is to collect bile (a yellowish-green fluid) produced in the liver and to carry the bile via the cystic duct to the gallbladder for storage and via the distal extrahepatic bile duct and through the pancreas to the small intestine for fat digestion.

Tumors of the intrahepatic bile ducts (e.g., ICC) arise from the biliary epithelium of small intrahepatic ductules or large intrahepatic ducts near the bifurcation of the right and left hepatic ducts.

Tumors of the perihilar (hilum) region (e.g., PCC or Klatskin tumor) originate from the main ducts of the hilum where the right and left hepatic ducts exit the liver and join to form the common hepatic duct.

Tumors of the extrahepatic bile ducts (e.g., ECC) occur in the common bile duct located between the point where the cystic duct joins the common hepatic duct and the ampulla of Vater.

CC generally evolves from four types of precursor lesions: (i) flat (biliary intraductal neoplasia [BilIN]), (ii) papillary (intraductal papillary neoplasm of the bile duct [IPNB]), (iii) tubular (intraductal tubular neoplasm of the bile [ITNB]), and (iv) mucinous cystic (hepatobiliary mucinous cystic neoplasia [hbMCN]) types [1,2].

BilIN contains epithelial cells with multilayering of nuclei and micropapillary projections that occur in the setting of chronic inflammation almost exclusively in the large bile ducts, giving rise to extrahepatic CC. IPNB displays papillary or villous excrescences (including biliary papilloma and papillomatosis) that occur preferentially within the intrahepatic bile ducts, leading to intrahepatic CC. ITNB is a lesion closely related to IPNB, showing a tubular rather than papillary growth. hbMCN is a cyst-forming tumor, usually without communication with bile ducts, with an ovarian-like subepithelial stroma, lined by mucinous cells. All these precursor lesions may demonstrate varying degrees of cytological atypia (increased nuclear to cytoplasmic ratio, loss of nuclear polarity, nuclear hyperchromasia) [1,2].

21.3 Epidemiology

CC is a rare disease with annual incidence of 0.35–1.8 per 100,000 in Europe, 0.6–1.0 per 100,000 in the United States, but as high as 71.3 per 100,000 in Asia (especially Thailand, China, and Korea). CC usually affects patients in the fifth to seventh decades of life, with a male-to-female ratio of 1.2–1.5:1.

CC in Thailand reflects the impact of chronic infection with liver fluke (*Opisthorchis viverrini*), while that in China and Korea relates to another closely related liver fluke (*Clonorchis sinensis*) [3].

21.4 Pathogenesis

Most cases of CC are sporadic without identifiable risk factors, whereas others are related to primary sclerosing cholangitis, bile duct cysts, choledocholithiasis, hepatolithiasis, liver fluke infection (*C. sinensis* and *O. viverrini*), Caroli disease, and Thorotrast [3].

CC is associated with mutational events at oncogenes (e.g., *KRAS*), tumor suppressor genes (e.g., *TP53, p16,* and *SMAD4*), chromatin-modifying genes (e.g., *ARID1A, BAP1,* and *PBMR1*), and *BRAF, PTEN, IDH1/IDH2* genes. Alterations involving components of oncogenic pathways (e.g., MEK, PI3K/AKT/mTOR, and PI3K/PI3KCA), tyrosine kinases (e.g., JAK/STAT3, ERK1/2, and p38MAPK), and hormone receptors (e.g., EGFR, HER2, MET, and VEGF) are also implicated [3].

Indeed, KRAS mutations are observed in PCC (22–53%) and ICC (9–17%); while IDH1/2 mutations are often detected in ICC. BAP1 and IDH1/2 mutations commonly occur in *Clonorchis sinensis*-related as well as non-liver fluke related CC; and TP53, KRAS, SMAD4, MLL3, ROBO2, RNF43 and PEG3 mutations are present in *Opisthorchis viverrini*-related CC. Gene rearrangements leading to oncogenic fibroblast growth factor receptor 2 (FGFR2) fusion proteins are noted in ICC (45%).

In infection-related CC, proinflammatory cytokines (e.g., interleukin-6 [IL-6]) activate inducible nitric oxide synthase, resulting in excess nitric oxide that mediates oxidative DNA-damage, inhibition of DNA repair enzymes, expression of cyclooxygenase 2 (COX-2), and ultimately cholestasis.

Interestingly, nongenomic upregulation of EGFR, human epidermal growth factor receptor 2 (HER2) and MET is linked to poor outcomes in CC. Further, BAP1 and PBMR1 mutations imply bone metastases and poor survival in ECC, whereas KRAS and TP53 mutations, as well as loss of PTEN expression, suggest poor survival in ICC [3].

21.5 Clinical features

Clinical symptoms of CC are due mainly to biliary tract obstruction and may range from jaundice (yellowing of the eyes and skin, especially ECC), fatigue, night sweats, cachexia, weight loss (30%–50%), abdominal pain (30%–50%), fever (20%), pruritus (66%), abnormal liver function tests (elevated bilirubin, alkaline phosphatase, gamma glutamyl transferase levels, carbohydrate antigen 19-9 [CA 19-9], and relatively normal transaminase levels), changes in stool or urine color, to death (usually within a year of diagnosis).

21.6 Diagnosis

Preoperative diagnosis of CC involves laboratory tests (e.g., liver function tests) and radiographic imaging procedures (e.g., abdominal ultrasound, CT, MRI, endoscopic retrograde cholangiography [ERC], magnetic resonance tomography [MRT] with cholangiography, PET-CT, or percutaneous transhepatic cholangiography [PTC]). A combination of MRT, CT, PET-CT, ERC, and PTC imaging techniques with histological examination allows determination of CC stages using the TNM system. In general, localized CCs are of Grade I, whereas disseminated CCs are of Grades II–IV [4].

Macroscopically, CC is a small, firm, grayish-white mass of 7–10 cm or multiple nodules of <1 cm with fibrous stroma. The tumor may be classified as mass-forming (ICC), periductal-infiltrating, or intraductal-papillary (PCC), according to its macroscopic growth pattern.

Histologically, ICC is an invasive, well-differentiated adenocarcinoma composed of neoplastic cells arranged in tubules, acini, and micropapillae in desmoplastic and inflammatory background. In the small bile duct type, small-sized tubules and acini invade the liver parenchyma. In the large bile duct type, the tumor mainly involves the intrahepatic large bile ducts. In the bile ductular type, neoplastic glands strongly resemble reactive ductules.

In the ductal plate malformation type, the tumors show minimal deviation features, reminiscent of congenital hepatic fibrosis or Caroli disease [2].

PCC and DCC are moderately to poorly differentiated adenocarcinomas with glandular and tubular structures, mucin production, and dense desmoplasia. In the papillary type, the lesion grows intraluminally and contains papillary excrescence of neoplastic cells, infiltrating foci with ordinary tubular pattern, and a mucinous component. In the tubular type, the exophytic lesion consists of tubules rather than papillae, and the infiltrating component is identical to that of ICC. In the superficial spreading type, neoplastic cells spread on the luminal surface of the bile ducts with only occasional infiltration of the wall [2].

CC is positive for mucin, CEA (cytoplasmic and luminal, not canalicular), CAM 5.2, AE1-AE3, keratin 903 (74%), CK7 (90%–96%) and CK19 (84%); reduced CK903, CK20 (30%–70%, often in nonperipheral tumors), epithelial membrane antigen, amylase, parathyroid hormone (PTH)-related peptide, p53 (10%–94%), proliferating cell nuclear antigen (PCNA), MOC31, and BerEP4, MUC1, MUC4, CD151, p27, COX2, fascin, tenascin, metallothionine, MMP7 may be a favorable prognostic factor; but negative for AFP and HepPar1. Molecularly, CC shows *KRAS* mutation, downregulation of β-catenin, and aberrant expression and activation of ErbB, hepatocyte growth factor (HGF), and IL6 [5].

Possible differential diagnoses for CC are gallstones, benign or malignant pancreatic tumors, sclerosing cholangitis, epithelioid hemangioendothelioma (mucin–), hepatocellular carcinoma (ductular differentiation) and metastatic adenocarcinoma (CK7–/CK20+).

21.7 Treatment

For localized (resectable) ICC and ECC (usually in the distal common bile duct), complete resection with negative surgical margins is effective [6].

For disseminated (unresectable) ICC, PCC, and DCC (which may invade directly into the portal vein, the adjacent liver, along the common bile duct, and to adjacent lymph nodes), complete surgical removal is impossible. Further, up to 64% patients with disseminated CC develop recurrent disease within 2–3 years postresection. Therefore, management of patients with disseminated CC or recurrent diseases involves locoregional treatments, systemic chemotherapy (e.g., gemcitabine cisplatin, capecitabine, oxaliplatin, leucovorin, irinotecan, and 5-fluorouracil), and symptomatic control [5].

21.8 Prognosis

Patients with localized (resectable) CC in the distal or middle portion of the extrahepatic bile duct have a better prognosis than those in the proximal third. Even with complete resection, ICC has a 5-year overall survival rate of 21%–63%, PCC 30%–40%, and DCC 20%–54%.

For disseminated (unresectable) CCs, the median overall survival is 37 months; the median progression-free survival is 9 months. The response rates are 62%, 33%, and 5% for partial response, stable disease, and progressive disease, respectively; and the 5-year survival rate is <5%.

References

1. Blechacz B. Cholangiocarcinoma: Current knowledge and new developments. *Gut Liver.* 2017; 11(1): 13–26.
2. Ebata T, Ercolani G, Alvaro D, Ribero D, Di Tommaso L, Valle JW. Current status on cholangiocarcinoma and gallbladder cancer. *Liver Cancer.* 2016; 6(1): 59–65.
3. Bridgewater JA, Goodman KA, Kalyan A, Mulcahy MF. Biliary tract cancer: Epidemiology, radiotherapy, and molecular profiling. *Am Soc Clin Oncol Educ Book.* 2016; 35: e194–203.
4. Bartella I, Dufour JF. Clinical diagnosis and staging of intrahepatic cholangiocarcinoma. *J Gastrointestin Liver Dis.* 2015; 24(4): 481–9.
5. Pathologyoutlines.com website. Cholangiocarcinoma (intrahepatic/peripheral). http://www.pathologyoutlines.com/topic/livertumorcholangiocarcinoma.html; accessed January 15, 2017.
6. PDQ Adult Treatment Editorial Board. *Bile Duct Cancer (Cholangiocarcinoma) Treatment (PDQ®): Health Professional Version.* PDQ Cancer Information Summaries. Bethesda, MD: National Cancer Institute (US); 2002–2016.

22
Colorectal Cancer

Kesara C. Ratnatunga, Jayantha Balawardena,
and Kemal I. Deen

22.1 Definition

According to the 2010 *WHO Classification of Tumours of the Digestive System*, tumors affecting the colon and rectum are classified into four categories: epithelial tumors, mesenchymal tumors, lymphomas, and secondary tumors.

Epithelial tumors of the colon and rectum are further separated into five groups: premalignant lesions, serrated lesions, hamartomas, carcinomas, and neuroendocrine neoplasms. C In turn, carcinomas are subdivided into five types: adenocarcinoma, adenosquamous carcinoma, spindle cell carcinoma, squamous cell carcinoma, and undifferentiated carcinoma.

Adenocarcinoma of the colon and rectum (commonly known as *colorectal cancer*, CRC; or colloquially *bowel cancer*) accounts for 98% of colonic cancer and represents the most common and important neoplasm of the digestive system, the third most common cancer, and the fourth most common cause of cancer death.

22.2 Biology

The colon (or the large intestine) is a tube of about 160 cm in length and 6–8 cm in diameter that connects the small intestine with the rectum. The colon may be divided into five sections: the cecum (which joins the ileum of the small intestine), the ascending colon, the transverse colon, the descending colon, and the sigmoid colon.

The rectum links the large intestine to the anal canal. It measures about 12 cm in length, has a similar diameter (6 cm) to the sigmoid colon at its commencement but is dilated near its termination (the rectal ampulla).

The main function of the colon and rectum is to absorb water, electrolytes, and nutrients from intestinal contents. On plain abdominal radiographs, the colon and rectum are seen to be filled with air and some fecal material.

Structurally, the wall of the colon and rectum is composed of mucosa (including epithelium, lamina propria, and muscularis mucosa), submucosa, muscularis propria (externa), and serosa (perimuscular tissue in rectum).

Colorectal cancer originates from the regenerating crypts of the mucosa, when gene mutation disrupts the natural process of apoptosis leading to the generation of immortal cell lines, causing heaps of cells that manifest as polyps. Over a period of approximately 5–10 years, further mutations in key genes (oncogenes and proto-oncogenes) result in the development of cells with potential for malignant change. Once colorectal cancer becomes established in the mucosa, it spreads loco-regionally and to remote locations, predominantly the liver and the lungs. Circumferential spread of tumor is associated with invasion of deeper tissues of the colorectal wall and beyond, to adjacent organs such as the cervix and uterus in women and the prostate gland and base of the urinary bladder in men. Further spread to lymph nodes along the inferior mesenteric artery and para-aortic region occurs via the lymphatics, whereas metastatic spread occurs via the venous blood stream predominantly to the liver and lungs, as well as the spine, spinal cord, and brain [1,2].

22.3 Epidemiology

The incidence of colorectal cancer currently stands at 40 per 100,000, a significant drop from 60.5 per 100,000 in 1970s. However, there has been a notable increase of colorectal cancer in patients younger than 50 years of age, chiefly due to a rise in the incidence of rectal cancer. Alarmingly, data predict that by the year 2030 the incidence of rectal cancer will increase by 124.2% in the 20–34 year age group and by 46% in the 35–49 year age group.

Sporadic colorectal cancer (60%) usually affects the older age groups (>50 years), predominantly in the rectum and left colon. Familial cancer (30%) occurs at an early age (mean age 45 years) and predominantly in the proximal colon more than in the rectum [2].

22.4 Pathogenesis

Risk factors for colorectal cancer include low-fiber-containing diets, diets with high levels of saturated animal fats, red meat, and processed meat; alcohol intake (in excess of 30 g daily), chronic inflammation (e.g., inflammatory bowel disease, ulcerative colitis, Crohn's disease), genetic mutations (e.g., *APC, KRAS*, DNA mismatch repair gene), and familial syndromes (e.g., Lynch syndrome) [1].

Mutation in the tumor suppressor gene APC (adenomatous polyposis coli) causes cells with unchecked cell division to occur at the surface of the crypt. A series of mutations in KRAS (Kirsten Rat Sarcoma) gene and then P53 gene abrogate tumor suppressor mechanisms and promote colorectal cancer development. If the initial mutation is inherited, the germ line cell will be affected at birth, with early appearance of polyps and malignant change. Familial adenomatous polyposis coli (FAP) represents such an example of inherited mutation in the APC gene, that presents with hundreds of colonic polyps in the adolescent years, and almost a 100% certainty of malignant transformation by age forty.

Mutation in the DNA mismatch repair gene disables the build-in mechanism for correcting replication errors in oncogenes, contributing to malignancy. Such mutation is observed in over 90% of patients with hereditary non polyposis colorectal cancer (HNPCC) and in up to 15% of sporadic colorectal cancer.

Chronic inflammation associated with inflammatory bowel disease (IBD), e.g., ulcerative colitis (UC) and Crohn's disease, can also induce genetic alterations leading to dysplasia and carcinoma. The presence of dysplasia heralds a 30% risk of colorectal cancer [2].

22.5 Clinical features

Clinical symptoms of colorectal cancer include (i) a change in bowel habit (e.g., diarrhea, constipation, or smaller, more frequent bowel movements), (ii) a change in appearance of stools (e.g., narrower stools or mucus in stools), (iii) a feeling of fullness or bloating even after a bowel movement, (iv) blood in the stools, (v) persistent abdominal (belly) pain or swelling, (vi) weakness or fatigue (due to anemia), (vii) unexplained weight loss, and (viii) a lump in the rectum or anus.

In general, right-sided tumors are associated with anemia, weakness, and fatigue; left-sided tumors cause change in bowel habits (diarrhea or constipation). Alteration of bowel habit, incomplete evacuation, tenesmus, and mucoid discharge or diarrhea often suggests a villous component or signet-ring tumor. Local spread of colorectal cancer anteriorly into the bladder base or prostate in men can cause hematuria or, in women, vaginal bleeding. Infiltration or compression of the ureters at the bladder neck will cause proximal dilation and eventually obstructive uropathy, which may remain asymptomatic until an advanced stage of the disease [3].

Involvement of the liver by metastasis may remain asymptomatic or cause right hypochondrial pain and even jaundice at a late stage; involvement of the lungs may result in respiratory symptoms; and involvement of spine and pelvic nerve may be associated with back pain and pelvic pain.

22.6 Diagnosis

Diagnosis of colorectal cancer involves physical examination (assisted by rigid proctoscopy, flexible endoscopy, colonoscopy), imaging (e.g., CT, MRI, PET), and histologic, immunological, and molecular examination of biopsy and tumor tissues [2].

Macroscopically, colorectal cancer is usually a single, polypoid, or ulcerated mass of 2–5 cm in size, which may cause serosal puckering if muscularis propria is involved. While right colon tumors are often polypoid and exophytic, left colon tumors tend to be annular, encircling lesions. Histologically, colorectal cancer shows well (15%–20% of cases), moderately (60%–70% of cases), or poorly (15%–20% of cases) differentiated gland-forming carcinoma with marked desmoplasia; glands filled with necrotic debris (dirty necrosis); inflammatory cells and scattered neuroendocrine cells; and intramural venous invasion. Six subtypes (i.e., cribriform comedo-type adenocarcinoma, medullary carcinoma, micropapillary carcinoma, mucinous adenocarcinoma, serrated adenocarcinoma, and signet-ring cell carcinoma) are recognized among adenocarcinoma affecting the colon and rectum. Immunohistochemically, the tumor stains positive for CK20, CDX2 (superior to villin), AMACR, estrogen receptor (occasionally), and CD10 (stromal cells) but negative for CK7 (except in rectal adenocarcinomas). At the molecular level, colorectal cancer often harbors mutations at the APC, TP53, and KRAS genes and shows microsatellite instability (MLH1, MSH2, MSH6, and PMS2) [2].

The staging of colorectal cancer is mostly based on the tumor–node–metastasis (TNM) system designed by the American Joint Committee on Cancer. The TNM system takes into account the various layers of local and regional tumor spread, number of nodes involved in assessment of nodal spread, and the organs involved in metastatic spread. Consequently, rectal cancer is classified into Stages I, IIA, IIB, IIIA, IIIB, IIIC, IV (Table 22.1). This staging information is central in determining the therapeutic approach and assessing prognosis [2,3].

Table 22.1 Staging of Rectal Cancer by the TNM System

Stage	T Stage	N Stage	M Stage
Stage I	T1, T2	N0	M0
Stage IIA	T3	N0	M0
Stage IIB	T4a	N0	M0
Stage IIC	T4b	N0	M0
Stage IIIA	T1,T2	N1/Nc	M0
	T1	N2a	M0
Stage IIIB	T3, T4a	N1/Nc	M0
	T2, T3	N2a	M0
	T1,T2	N2b	M0
Stage IIIC	T4a	N2a	M0
	T3, T4a	N2b	M0
	T4b	N1, N2	M0
Stage IVA	Any T	Any N	M1a
Stage IVB	Any T	Any N	M1b

Note: **Tumor stage**

T1- Tumor invades the submucosa

T2- Tumor invades the muscularis propria

T3- Tumor invades through the muscularis propria into the peri rectal tissue

T4a- Tumor penetrates through to surface of the visceral peritoneum

T4b- Tumor invades or adheres to adjacent organs or structures

Lymph node stage

N0- No regional node metastasis

N1- Metastasis in 1-3 regional lymph nodes

 N1a- One lymph node involved

 N1b- 2-3 lymph nodes involved

 N1c- Tumor deposit(s) in the mesentery, subserosa or non-peritonealized perirectal tissue with no lymph node involvement

N2- Metastasis in 4 or more regional lymph nodes

 N2a- 4-6 lymph nodes involved

 N2b- 7 or more lymph nodes involved

Distal organ involvement

M0- No distal metastasis

M1- Distal metastasis

 M1a- Single organ or site involved

 M1b- Metastasis in more than one organ/site or peritoneum

22.7 Treatment

Surgery remains the mainstay of cure in colorectal cancer. Chemotherapy and radiotherapy play a critical role in the adjuvant as well as neoadjuvant settings. Newer therapies such as targeted therapy have a place in specific situations [4,5].

The surgical resection and approach depends on the location of the tumor, its local spread, and respectability. Preoperative irradiation has certain advantages, as it deals with tissue that is not violated by surgery, which is better vascularized, better oxygenated, and more responsive to irradiation therapy.

Adjuvant chemotherapy includes 6–12 cycles of folinic acid, 5-fluorouracil, and oxaliplatin (FOLFOX)-based regime or capecitabine and oxaliplatin (CapeOx) as first-line therapy. Folinic acid 5-fluorouracil and irinotecan (FOLFIRI) may be used as an alternative to oxaliplatin with similar efficacy.

Neoadjuvant therapy may be given as either long course chemoradiation (over 5 weeks) or as short course irradiation alone. Tattooing the lesion prior to neoadjuvant therapy is important, such that, if the lesion shows a complete clinical response, the tattoo mark will serve as a guide to subsequent resection to achieve a satisfactory resection margin.

22.8 Prognosis

The 5-year survival rates for colorectal cancer Stages I, IIA, IIB, IIIA, IIIB, IIIC, and IV are estimated to be 87%, 80%, 49%, 84%, 71%, 58%, and 12%, respectively.

References

1. Kheirelseid EAH, Miller N, Kerin MJ. Molecular biology of colorectal cancer: Review of the literature. *Am J Mol Biol.* 2013; 3: 72–80.
2. National Comprehensive Cancer Network (NCCN). *Rectal Cancer Clinical Practice Guideline Version 2*; NCCN, Fort Washington, PA, 2016.
3. Steele S, Hull T, Read T, Saclarides T, Senagore A, Whitlow C. *The ASCRS Textbook of Colon and Rectal Surgery*; 2016.
4. PDQ Adult Treatment Editorial Board. *Colon Cancer Treatment (PDQ®): Health Professional Version.* PDQ Cancer Information Summaries. Bethesda, MD: National Cancer Institute (US); 2002–2016.
5. PDQ Adult Treatment Editorial Board. *Rectal Cancer Treatment (PDQ®): Health Professional Version.* PDQ Cancer Information Summaries. Bethesda, MD: National Cancer Institute (US); 2002–2016.

23
Esophageal Cancer

Megan M. Boniface

23.1 Definition

The esophagus is affected predominantly by two histologic types of cancer: esophageal squamous cell carcinoma (ESCC) and esophageal adenocarcinoma (EAC) (Table 23.1). ESCC typically develops in the mid-esophagus, whereas EAC is usually found in the distal esophagus. These two types account for nearly 95% of esophageal cancers, although sarcomas, small cell carcinomas, lymphomas, and carcinoids may occur in the esophagus as well.[1,2]

23.2 Biology

Situated posteriorly to the trachea in the neck and thoracic regions, the esophagus (commonly known as the *food pipe* or *gullet*) is a thin, muscular tube of 25 cm in length and 2 cm in diameter that connects the pharynx (throat) to the stomach.

Structurally, the esophageal wall from the lumen outwards consists of mucosa (nonkeratinized stratified squamous epithelium, lamina propria, and muscularis mucosae), submucosa (connective tissue), muscularis layer (striated and smooth muscle), and adventitia layer (loose connective tissue). The esophagus has two muscular rings (or sphincters) in its wall, one at the top and one at the bottom.

Tumors of the esophagus mostly arise from the squamous or columnar epithelium of the mucosa.

23.3 Epidemiology

Over 450,000 people are diagnosed with esophageal cancer, making it the eighth most common malignancy worldwide. The risk for developing esophageal cancer increases with age and shows a preponderance for men over women. Incidence varies by both histologic type and geographic location with the majority of cases attributed to ESCC, and the highest rates found in eastern Asia and eastern and southern Africa. The "esophageal

Table 23.1   Comparison of ESCC and EAC

	ESCC	EAC
Gender	Male	Male
Race	African-American	Caucasian
Median age, years	62.7	53.4
Typical location	Mid-esophagus	Distal esophagus
Environmental factors	Alcohol, smoking, drinking hot liquids, poor nutritional status	Obesity, chronic GERD, smoking

Note: EAC, esophageal adenocarcinoma; ESCC, esophageal squamous cell carcinoma; GERD, gastroesophageal reflux disease.

cancer belt" identifies the highest-risk area extending from northeast China to the Middle East, with >90% of cases in this region being ESCC and affecting both genders equally.[3,4]

In contrast, EAC is the most common type of esophageal cancer in the Western world, where ESCC predominated 30 years ago. This change probably reflects a decrease in risk factors for ESCC such as smoking and alcohol use and an increase in obesity and subsequent diagnoses such as gastroesophageal reflux disease (GERD) and Barrett's esophagus.[2,4]

23.4 Pathogenesis

Squamous cell carcinoma. Risk factors for ESCC differ in high-risk and low-risk areas. In high-risk areas and the esophageal cancer belt, predisposing factors seem to involve poor diet and nutritional status and drinking beverages at hot temperatures. In lower-risk areas such as Western countries and the United States, tobacco and alcohol abuse account for almost 90% of cases.[4]

Adenocarcinoma. Risk factors for EAC include obesity, chronic GERD, Barrett's esophagus, smoking history, and diets low in fruits and vegetables. The majority of EAC appear in the distal esophagus and arise from Barrett's esophagus, which is a condition that occurs as a result of chronic GERD, where columnar epithelium replaces the squamous epithelium that normally lines the esophagus. The shift in predominate histologic type of esophageal cancer in the Western world over the past several decades to EAC is attributed to the rise in obesity and subsequent effects such as GERD and Barrett's esophagus. Table 23.1 depicts differences in risk factors and tumor location between ESCC and EAC.[1,3]

23.5 Clinical features

Dysphagia is the most common presenting symptom of esophageal cancer. Patients may also present with weight loss, reflux/regurgitation, or retrosternal chest discomfort. Other possible symptoms include hoarseness (associated with laryngeal nerve involvement), hiccups, cough, pneumonia, and gastrointestinal bleeding.[3]

23.6 Diagnosis

Patients suspected of esophageal cancer should be examined by esophagogastroduodenoscopy to allow for endoscopic visualization and biopsy. The majority of ESCC arise from the middle and upper third of the esophagus, whereas EAC typically develops in the distal esophagus.[1] Once a diagnosis of esophageal cancer is established with tissue specimen, a preoperative staging evaluation is performed and tumor is then classified using the universal tumor–node–metastasis (TNM) staging criteria of the American Joint Committee on Cancer in order to determine the appropriate treatment pathway.[5] Computed tomography (CT) is performed first to help ascertain the extent of local disease (N stage) and evaluate for distant metastases (M stage). If there is no evidence for distant spread on CT, F-fluoro-2-deoxy-D-glucose position emission tomography with computed tomography (FDG-PET-CT) can be used as a more sensitive tool for the detection of metastatic disease based on altered glucose metabolism. This is essential in determining whether treatment will take a curative or palliative approach.[6]

If no evidence for distant metastases is found on imaging, endoscopic ultrasound (EUS) is used to determine the depth of tumor invasion into the esophageal wall (T stage) and to examine for the presence of regional lymph nodes (N stage). EUS demonstrates an accuracy of 73% to 89% for T staging, and 90% for assessing nodal involvement. In addition, EUS has the highest sensitivity for detecting regional lymph node metastases using EUS-guided fine-needle aspiration to sample.[3,6]

23.7 Treatment

Treatment selection for esophageal cancer is based on TNM staging criteria.[7]

Early esophageal cancer. Due to increased screening and routine endoscopic surveillance for Barrett's esophagus, the incidence of superficial esophageal cancer is rising. Superficial esophageal cancers are defined as tumors that invade no deeper than the submucosa. Accurate assessment of the depth of invasion (T stage) is critical in determining the extent of disease and treatment selection. T1a tumors are defined as intramucosal tumors and T1b tumors invade into the submucosa. Depending on the extent of submucosal involvement, treatment ranges from endoscopic resection +/− ablation to surgical esophagectomy. This drastic difference in treatment correlates with the risk of lymph node metastases. In one series of close to 4,000 patients derived from the National Cancer Database, the risk of lymph node metastases was 5% for T1a tumors versus 17% for T1b tumors.[8]

Other factors that may attribute to the risk of lymph node metastases include tumor size, macroscopic appearance, histologic differentiation, and the presence of lymphovascular invasion. In some series, tumors >2 cm, non-flat-type lesions, intermediate-/high-grade lesions, and those with lymphovascular invasion were found to have a higher rate for nodal metastases.[8,9] Therefore, patients with intramucosal (T1a) tumors without these concerning features may be treated with endoscopically and patients with submucosal tumors (T1b) or intramucosal tumors with concerning features should undergo esophagectomy.[8]

Locoregional esophageal cancer. Patients with tumors that are localized to the esophagus and/or regional lymph nodes without evidence for distant metastatic disease (cT1bN+, cT2-T4aN0-N+) should undergo neoadjuvant chemoradiation followed by surgical resection. This recommendation is the result of several trials that demonstrated a significant survival benefit for patients that undergo preoperative chemoradiotherapy followed by surgery compared with surgery alone. Following the CROSS trial and National Comprehensive Cancer Network guidelines, the neoadjuvant chemotherapy regimen typically used is carboplatin and paclitaxel in combination with concurrent radiation therapy (41.4–50.4 Gy). This recommendation holds true for both histologies, except for patients with ESCC of the cervical esophagus or those who are not surgical candidates or decline surgery. These patients are typically treated with definitive chemoradiation with a higher radiation dose of 50–50.4 Gy.[7] After the completion of neoadjuvant treatment, repeat imaging is performed to evaluate for treatment response and rule out metastatic disease. Patients that remain resectable based on repeat imaging are taken for esophagectomy.[10]

Metastatic and unresectable esophageal cancer. Patients with esophageal cancer invading into adjacent structures (heart, great vessels, and trachea) or adjacent organs and those with distant metastatic disease are unresectable and, therefore, incurable.[5] Treatment in this group of patients is aimed at symptom palliation and improving quality of life. Depending on performance status and comorbidities, treatment options consist of systemic chemotherapy alone or in conjunction with local therapies (radiation or stent placement) to attempt to manage obstructive symptoms.[3] Tumors should be assayed for HER2-neu overexpression and trastuzumab should be added for HER-2-positive tumors if there are no contraindications for use. Palliative measures such as esophageal stent or feeding tube placement may be necessary in patients with completely obstructive tumors or poor nutritional status.[7]

23.8 Prognosis

The overall 5-year survival for esophageal cancer ranges from 15% to 25% with better outcomes in patients diagnosed with earlier stage of disease.[3] These dismal numbers are associated with the fact that >30% of patients are found to have metastatic disease at the time of diagnosis.[1] Prognosis strongly correlates to stage of disease. In a study by Ell and colleagues, endoscopic mucosal resection (EMR) and photodynamic therapy was used to treat T1a intramucosal cancers with a reported estimated 3-year survival of 98%.[3] By comparison, data from the SEER database from 1998–2009 show that patients with metastatic esophageal cancer have an overall 5-year survival of 3.4%.[1]

References

1. Zhang Y. Epidemiology of esophageal cancer. *World J Gastroenterol.* 2013;19(34):5598–606.
2. Eslick GD. Epidemiology of esophageal cancer. *Gastroenterol Clin North Am.* 2009;38(1):17–25, vii.
3. Pennathur A, Gibson MK, Jobe BA, et al. Oesophageal carcinoma. *Lancet.* 2013;381(9864):400–12.
4. Torre LA, Bray F, Siegel RL, et al. Global cancer statistics, 2012. *CA Cancer J Clin.* 2015;65(2):87–108.
5. AJCC. *Cancer Staging Manual.* 7th edition. New York: Springer; 2010.
6. van Vliet EP, Heijenbrok-Kal MH, Hunink MG, et al. Staging investigations for oesophageal cancer: A meta-analysis. *Br J Cancer.* 2008;98(3):547–57.
7. Network NCC. *NCCN Clinical Practice Guidelines in Oncology— Esophageal and Esophagogastric Junction Cancers (Version 3.2015).* http://www.nccn.org/professionals/physician_gls/pdf/esophageal. pdf. Accessed September 29, 2016.

8. Merkow RP, Bilimoria KY, Keswani RN, et al. Treatment trends, risk of lymph node metastasis, and outcomes for localized esophageal cancer. *J Natl Cancer Inst.* 2014;106(7). pii: dju133.

9. Shimada H, Nabeya Y, Matsubara H, et al. Prediction of lymph node status in patients with superficial esophageal carcinoma: analysis of 160 surgically resected cancers. *Am J Surg.* 2006;191(2):250–4.

10. van Hagen P, Hulshof MC, van Lanschot JJ, et al. Preoperative chemo-radiotherapy for esophageal or junctional cancer. *N Engl J Med.* 2012;366(22):2074–84.

24
Gallbladder Cancer

24.1 Definition

Like other organs in the digestive system, the gallbladder is susceptible to primary epithelial tumors (e.g., carcinoma *in situ*, adenocarcinoma, squamous cell carcinoma [SCC], adenosquamous carcinoma), mesenchymal tumors (e.g., sarcoma), and lymphoma, as well as secondary tumors.

The most common histologic tumor type affecting the gallbladder is adenocarcinoma, which consists of three subtypes: nonpapillary adenocarcinoma (75%), papillary adenocarcinoma (5%), and mucinous adenocarcinoma (2.5%). Altogether these adenocarcinoma subtypes account for 82% of gallbladder malignancies reported, whereas SCC, adenosquamous carcinoma, small cell carcinoma, signet-ring cell carcinoma, neuroendocrine tumor, and sarcoma are responsible for the rest [1].

24.2 Biology

The gallbladder is a pear-shaped sac of 10 cm in size and 30–50 ml in capacity that sits just under the liver. It can be divided into three parts: (i) fundus (the rounded, end portion projecting into the inferior surface of the liver; (ii) body (the main portion occasionally in contact posteriorly with the transverse colon and proximal duodenum); and (iii) neck (the tapering portion connecting with the cystic duct and showing susceptibility to gallstones).

Structurally, the gallbladder is composed of mucosa (with mucus-secreting columnar epithelium), tunica (lamina) propria, muscularis propria (crisscrossing muscles), subserosa (connective tissue), and serosa (on the inferior peritoneal surface only, not on the superior hepatic surface). There is a notable absence of submucosa in the gallbladder. The cystic duct contains spiral folds of mucosa (called the *valves of Heister*).

The main functions of the gallbladder are to store the bile produced by the liver while not eating and to supply the bile to the small intestine for fat digestion during meals.

Adenocarcinoma arises in the gland cells lining the gallbladder, with papillary adenocarcinoma developing in the connective tissues that hold the gallbladder in place, and mucinous adenocarcinoma found in pools of mucus. SCC evolves from the skin-like cells along with the gland cells lining the gallbladder. Adenosquamous carcinoma has both squamous cell and glandular cell components. Small cell (oat cell) carcinoma contains abnormal cells of a distinctive oat shape. Sarcoma begins in the muscle layer of the gallbladder [2].

24.3 Epidemiology

Gallbladder cancer has annual incidence rates of 0.4–1.4 cases per 100,000 in Denmark, Norway, UK, United States, Canada, Australia, and New Zealand but between 4.7 and 27 cases per 100,000 in Asia, South America, and other parts of the world. People of 50–60 years and 70–80 years of age are frequently affected, and women seem to be highly vulnerable (with a female-to-male ratio of 2.5:1) [3].

24.4 Pathogenesis

Risk factors for gallbladder cancer include obesity, gallbladder abnormalities (cholelithiasis/gallstone, porcelain gallbladder, gallbladder polyps, congenital biliary cysts, and pancreaticobiliary maljunction anomalies), exposures (heavy metals, methyldopa, isoniazid, estrogen, and smoking), chronic infection (e.g., *Salmonella* typhi and paratyphi, and *Helicobacter*), and genetic alterations (e.g., mutations at *TP53, P16, KRAS, BRAF, EGFR,* and *PI3K/PIK3CA* genes; loss of heterozygosity at chromosomes 1p34–36 [p73], 3p [VHL, RAR-beta, RASSF1A, and FHIT], 5q21 [APC], 8p21–23 [PRLTS and FEZ1], 9p21 [p15, p16], 9q [DBCCR1], 13q14 [RB], 16q24 [WWOX and FRA16D], 17p13 [p53], 18q, and 22q; and overexpression of EGFR, HER2, MET, and VEGF) [3].

24.5 Clinical Features

Symptoms of gallbladder cancer usually appear at the later stages of disease and may include pain above the stomach (54%–83%), jaundice (10%–46%), anorexia (4%–41%), nausea/vomiting (15%–43%), weight loss (10%–39%), fever, bloating, and lumps in the abdomen (3%–8%).

24.6 Diagnosis

Diagnosis of gallbladder cancer is achieved through the use of laboratory tests (e.g., liver function tests, CA 19-9 assay, and carcinoembryonic antigen

[CEA] assay) and imaging studies (ultrasonography, CT, MRI, percutaneous transhepatic cholangiography, endoscopic retrograde cholangiography, and magnetic resonance cholangiopancreatography).

Macroscopically, gallbladder cancer may form a gray-white mass of 2.9–4.8 cm in size, localized wall thickening with induration of the wall, or polypoidal growth, leading to distension or collapse of the gallbladder. Histologically, gallbladder cancer may appear as nonpapillary adenocarcinoma (or simply adenocarcinoma; typically co-presence of gallstone), papillary adenocarcinoma (papillae with vascular core and minimal pleomorphism, often accompanied by mucin production), mucinous adenocarcinoma (single cells or clusters with >50% stromal mucin deposition, larger than adenocarcinoma 4.8 cm vs. 2.9 cm), signet-ring cell carcinoma (a predominance of signet-ring cells, and intracytoplasmic mucin displacing the nuclei to the periphery), adenosquamous carcinoma (a mixture of glandular and squamous components; co-presence of gallstone), SCC (atypical keratinized cells and/or polliwog cells in a necrotic background; co-presence of gallstone), neuroendocrine carcinoma (rosettes, salt/pepper chromatin, anisonucleosis, and/or nuclear molding), small cell carcinoma (smudge cells, scant cytoplasm, necrosis, salt/pepper chromatin, and/or nuclear molding; more bulky than adenocarcinoma), and undifferentiated carcinoma NOS (dispersed and highly pleomorphic cells with abundant necrosis) [3].

Immunohistochemically, gallbladder adenocarcinoma is positive for cytokeratin 7 (CK7), and focally positive for carcinoembryonic monoclonal antibody (CEA-M), CA19-9, MUC1, B72.3, and MUC5AC.

The stages of gallbladder cancer are determined on the basis of tumor invasion, involvement of nearby lymph nodes, and status of spread. They are classified as stage 0 (carcinoma in situ), I (localized tumor), II, IIIA, IIIB, IVA, and IVB (disseminated tumor).

24.7 Treatment

The treatments options for gallbladder cancer include surgery, radiation therapy, chemotherapy, and palliative therapy [4–6].

Localized tumor (Stage I), which represents a minority of cases and is confined to the gallbladder wall, can be completely resected. Regional lymphatics and lymph nodes should be removed along with the gallbladder in such patients [5].

Disseminated tumors (Stages II–IV), which represent the majority of cases and have invaded directly into adjacent liver or biliary lymph nodes or disseminated throughout the peritoneal cavity, cannot be completely resected. At thithese stages, standard therapy is directed at palliation including radiation therapy and radiosensitizer drugs [5].

24.8 Prognosis

Gallbladder cancer generally has a poor prognosis, as it typically remains silent until an advanced and often noncurative stage [6].

The early stage, localized gallbladder cancer (carcinoma *in situ*, or *mucosal lesion*) has overall 5-year survival rates of 75%–90% with surgical resection and stage-adjusted therapy.

The later-stage, disseminated gallbladder cancer has overall 5-year survival rates of 60% for Stage II (tumor growing into muscle tissue but not through serosa), 20%–25% for Stage III (tumor penetrating the serosa or invading the one adjacent organ), and <10% for Stage IV (tumor invading the main portal vein, hepatic artery, or two or more adjacent organs), respectively. In fact, 85% of patients with Stages III and IV gallbladder cancer have an overall survival of only 2–8 months.

References

1. Ebata T, Ercolani G, Alvaro D, et al. Current status on cholangiocarcinoma and gallbladder cancer. *Liver Cancer*. 2016;6(1):59–65.
2. Kanthan R, Senger JL, Ahmed S, et al. Gallbladder cancer in the 21st century. *J Oncol*. 2015;2015:967472.
3. Bridgewater JA, Goodman KA, Kalyan A, et al. Biliary tract cancer: Epidemiology, radiotherapy, and molecular profiling. *Am Soc Clin Oncol Educ Book*. 2016;35:e194–203.
4. Aloia TA, Járufe N, Javle M, et al. Gallbladder cancer: Expert consensus statement. *HPB (Oxford)*. 2015;17(8):681–90.
5. PDQ Adult Treatment Editorial Board. *Gallbladder Cancer Treatment (PDQ®): Health Professional Version*. PDQ Cancer Information Summaries. Bethesda, MD: National Cancer Institute (US); 2002–2016.
6. Goetze TO. Gallbladder carcinoma: Prognostic factors and therapeutic options. *World J Gastroenterol*. 2015;21(43):12211–7.

25
Gastric Cancer

25.1 Definition

Tumors of the stomach (also referred to as *stomach cancer or gastric cancer*) include *primary epithelial tumors* (intraepithelial neoplasia [adenoma], carcinoma [adenocarcinoma—intestinal/diffuse, papillary adenocarcinoma, tubular adenocarcinoma, mucinous adenocarcinoma, signet-ring cell carcinoma, adenosquamous carcinoma, squamous cell carcinoma, small cell carcinoma, and undifferentiated carcinoma], and carcinoid [well-differentiated endocrine neoplasm]); *primary mesenchymal tumors* (leiomyoma, schwannoma, granular cell tumor, glomus tumor, leiomyosarcoma, and GI stromal tumor—benign/uncertain malignant potential/malignant, Kaposi sarcoma); *primary malignant lymphomas* (marginal zone B-cell lymphoma of MALT-type, mantle cell lymphoma, diffuse large B-cell lymphoma); and *secondary tumors*.

Out of these the histological type of adenocarcinoma (encompassing intestinal and diffuse adenocarcinomas of Lauren classification; and tubular, papillary, and mucinous adenocarcinomas as well as signet-ring cell carcinoma of WHO classification) makes up about 90% of malignant tumors of the stomach, whereas MALT lymphoma, leiomyosarcoma, and other rare tumors account for the remainder.

Due to the fact that traditional histomorphologic classification of gastric tumors do not satisfactorily provide details relevant to clinical utilities and treatment guidelines, application of genetic techniques has led to the identification of four molecular subtypes: microsatellite stable (MSS)/epithelial-mesenchymal transition (EMT) subtype, microsatellite instable (MSI) subtype, MSS/tumor protein 53 (TP53)+ subtype, and MSS/TP53− subtype. It is notable that the MSS/EMT subtype has the highest frequency of recurrence (63%) and the worst prognosis; the MSI subtype has the lowest frequency of recurrence (22%) and the best overall prognosis. The MSS/TP53+ and MSS/TP53− subtypes have intermediate recurrence rates and prognosis. A more recent molecular classification system separates gastric tumors into four major subtypes: Epstein–Barr virus (EBV)-associated tumors, MSI tumors, tumors with chromosomal instability (CIN), and genomically stable (GS) tumors [1].

25.2 Biology

Forming part of the gastrointestinal tract, the stomach is a J-shaped bag located in the superior aspect of the abdomen. The stomach consists of four main regions: cardia (surrounding the superior opening of the stomach that connects to the esophagus), fundus (the rounded portion superior to and left of the cardia), corpus (the large central portion inferior to the fundus), and pylorus (connecting the stomach to the duodenum):

Structurally, the stomach wall is composed of five layers (from superficial to deep): serosa (peritoneum), subserosa, muscularis externa, submucosa, and mucosa. The serosa (peritoneum) covering most of the stomach helps attach it to the abdominal wall. The muscularis externa includes three layers of smooth muscles: inner oblique (unique to stomach), middle circular (forming the pylorus), and outer longitudinal. The submucosa comprises loose connective tissue, blood vessels, and Meissner's nerve plexus. The mucosa lines the stomach cavity and contains columnar epithelium (1 mm in height), lamina propria, and muscularis mucosa, covered by a 100–200 µm thick mucus. Gastric foveolae extend to the muscularis mucosa (where the tubular glands formed by different exocrine cells are located).

The columnar cells in the stomach secrete mucin; the chief (zymogenic) cells in the fundus secrete protein digesting pre-enzyme pepsinogen; the parietal (oxyntic) cells in the corpus (body) secrete acid (H^+ ions) and intrinsic factor; and the G cells in the antrum secrete gastrin (which in turn acts on parietal cells).

The main functions of the stomach are to store and digest ingested food for nutrient supply to the body and to eliminate most microbial pathogens with acidic gastric juice prior to their infiltration into the intestines and other abdominal organs.

Intestinal adenocarcinoma, with a distal location to the cardia, is a well-differentiated tumor that may have arisen from intestinal cells. It contains the glandular cells arranged in tubular or glandular structures on a background of intestinal metaplasia. The terms *tubular* (with prominent dilated or slit-like, branching tubules), *papillary* (with well-differentiated tumor cells and exophytic growth), and *mucinous* (with extracellular mucinous pools) are assigned to the various types of intestinal adenocarcinomas that closely relate to those of WHO classification.

Diffuse adenocarcinoma, with a proximal location to the cardia, is an undifferentiated or poorly differentiated tumor that appears to arise from the middle layer of mucosa. It has round, small cells, diffuse infiltration in the gastric wall, and few or no glands. It correlates to signet-ring cell carcinoma (which is defined as predominantly or exclusively signet-ring cells) of WHO classification.

25.3 Epidemiology

Gastric cancer is the fifth most common malignancy (behind lung, breast, and colorectal cancers) and the third leading cause of cancer-related deaths (10%) worldwide. The incidence rates range from 2 to 62.2 cases per 100,000, with 62.2, 48.2, 46.8, and 41.3 cases per 100,000 males in Korea, Mongolia, Japan, and China, respectively [2].

Intestinal adenocarcinoma affects people of about 55 years of age, with a male predominance (2/3 being men). It is commonly detected in developing countries, in black people, in lower socioeconomic groups, and in those with chronic infection by *Helicobacter pylori* or EBV.

Diffuse adenocarcinoma occurs in people of about 48 years of age, with an equal male-to-female ratio. It is often found in developed countries, in white people, in higher socioeconomic levels, and in those with gastro-esophageal reflux disease and obesity.

25.4 Pathogenesis

Risk factors for gastric cancer include diet (low in citrus fruits and green leafy vegetables; high in salted, smoked, or preserved foods), infection (e.g., *H. pylori* and EBV), advanced age, male gender, chronic atrophic gastritis, intestinal metaplasia, pernicious anemia, gastric adenomatous polyps, family history of gastric cancer, cigarette smoking, Ménétrier disease (giant hypertrophic gastritis), and familial adenomatous polyposis [3].

Occurring mainly in the gastric fundus or corpus and accounting for 10% of gastric cancer, EBV-associated tumors (or EBV-associated gastric cancer) demonstrate frequent DNA hypermethylation involving both promoter (e.g., cyclin-dependent kinase inhibitor 2A [*CDKN2A*] promoter) and nonpromoter CpG islands; somatic gene alterations in phosphatidylinositol-4,5-bisphosphate 3-kinase catalytic subunit alpha (*PIK3CA*), AT-rich interactive domain 1A (*ARID1A*), and BCL6 corepressor (*BCOR*), *TP53*, janus kinase 2 (*JAK2*), *CD274*, and programmed cell death 1 ligand 2 (*PDCD1LG2*) [1,4,5].

Representing 21% of gastric cancer, MSI tumors are associated with older age, female gender, intestinal type, and less aggressive tumor stages.

Besides containing mutations in *PIK3CA*, *ERBB3*, *HER2*, and epidermal growth factor receptor (*EGFR*), MSI subtype shows widespread replication errors in simple repetitive microsatellite sequences due to the defects in mismatch repair genes [1,4,5].

Frequently found in the gastroesophageal junction/cardia and representing 50% of gastric cancer, tumors with CIN relate to intestinal type histology and often have copy number gains at 8q, 12q, 13q, 17q, and 20q and copy number losses at 3p, 4q, 5q, 15q, 16q, and 17q. In addition, CIN subtype shows amplifications of genes encoding receptor tyrosine kinases, *EGFR*, cell cycle genes cyclin E1 (*CCNE1*), cyclin D1 (*CCND1*), and cyclin-dependent kinase 6 (*CDK6*) [1,4,5].

GS tumors represent 19.6% of gastric cancer and are enriched with diffuse histological variant. GS subtype often shows mutation in ras homolog family member A (*RHOA*) and cadherin 1 (*CDH1*), as well as interchromosomal translocation between claudin 18 (*CLDN18*) and Rho GTPase-activating protein 6 (*ARHGAP26*) [1,4,5].

25.5 Clinical features

Clinical symptoms of gastric cancer range from poor appetite, weight loss, abdominal (belly) pain, vague discomfort in the abdomen (usually above the navel), a sense of fullness in the upper abdomen after eating a small meal, heartburn, indigestion, nausea, vomiting (with or without blood), swelling or fluid buildup in the abdomen, to low red blood cell count (anemia) [6].

25.6 Diagnosis

Intestinal adenocarcinoma is a bulky tumor composed of glandular structures. Microscopically, the tumor contains apical mucin vacuoles, calcification, endocrine cells, and rare Paneth cells. Mucin-secreting columnar cells secrete are rarely ciliated, and tubules show branching, tortuous, anastomosing, and plexiform structures. The tumor stains positive for acid mucins (Alcian blue, colloidal iron) and p53 [6].

Diffuse adenocarcinoma (also called *poorly cohesive*, *linitis plastica*, or *signet-ring adenocarcinoma*) is characterized by extensive infiltration of poorly differentiated discohesive malignant cells over gastric wall, creating a thickened, rigid, leather bottle-like stomach and leading to pyloric obstruction. Microscopically, the tumor contain castric-type mucus cells (mucus-secreting cells), infiltrating (sometimes transmurally) as individual cells or small clusters; numerous signet-ring cells (mucin pushing nucleus to periphery), submucosal fibrosis, mucosal ulceration, hypertrophic muscularis propria, and marked desmoplastic and inflammatory reaction. The tumor stains positive for mucicarmine, Alcian blue-PAS, carcinoembryonic antigen, epithelial membrane antigen, keratin, and villin but negative for TTF1 and p53 [6].

The stages of gastric cancer are defined as 0, IA, IB, IIA, IIB, IIIA, IIIB, IIIC and IV by the TNM system of AJCC. The availability of this information is crucial for selecting appropriate treatments and predicting disease outcomes.

25.7 Treatment

Treatment options for gastric cancer include surgery (e.g., distal or proximal subtotal gastrectomy, total gastrectomy), chemotherapy, chemoradiation, radiotherapy, and targeted therapy (eg, anti-HER2 trastuzumab) [7,8].

For stage 0 gasric cancer confined to the mucosa, gastrectomy with lymphadenectomy leads to survival beyond 5 years in 90% of cases.

For stage I gastric cancer, surgical resection including regional lymphadenectomy is the treatment of choice. Postoperative chemoradiation therapy may be considered for with node-positive (T1 N1) and muscle-invasive (T2 N0) disease.

For stage II gastric cancer, surgical resection with regional lymphadenectomy is the treatment of choice. Postoperative chemoradiation therapy, as well as peri-/post-operative chemotherapy may be considered.

For stage III gastric cancer, surgery alone is curative in 15% of cases; postoperative chemoradiation therapy helps extend survival to 36 months vs. 27 months by surgery alone; and postoperative chemotherapy (e.g., capecitabine/oxaliplatin) gives a 3-year disease-free survival rate of 74% compared to 59% by surgery-alone.

For state IV gastric cancer, palliative chemotherapy, endoluminal laser therapy, endoluminal stent placement, gastrojejunostomy, palliative radiation therapy and palliative resection may be considered.

25.8 Prognosis

The 5-year overall survival rate for patients with localized distal gastric cancer is >50%; that for patients with proximal gastric cancer is 10%–15%; and that for patients with disseminated disease is 0%.

References

1. Chen T, Xu XY, Zhou PH. Emerging molecular classifications and therapeutic implications for gastric cancer. *Chin J Cancer.* 2016;35:49.
2. Ang TL, Fock KM. Clinical epidemiology of gastric cancer. *Singapore Med J.* 2014;55(12):621–8.
3. Garattini SK, Basile D, Cattaneo M, et al. Molecular classifications of gastric cancers: Novel insights and possible future applications. *World J Gastrointest Oncol.* 2017;9(5):194–208.
4. Choi YY, Noh SH, Cheong JH. Molecular dimensions of gastric cancer: Translational and clinical perspectives. *J Pathol Transl Med.* 2016;50(1):1–9.
5. Lim B, Kim JH, Kim M, Kim SY. Genomic and epigenomic heterogeneity in molecular subtypes of gastric cancer. *World J Gastroenterol.* 2016;22(3):1190–201.
6. Sung JK. Diagnosis and management of gastric dysplasia. *Korean J Intern Med.* 2016;31(2):201–9.
7. PDQ Adult Treatment Editorial Board. *Gastric Cancer Treatment (PDQ®): Health Professional Version.* PDQ Cancer Information Summaries. Bethesda, MD: National Cancer Institute (US); 2002–2016.
8. Smith JP, Nadella S, Osborne N. Gastrin and gastric cancer. *Cell Mol Gastroenterol Hepatol.* 2017;4(1):75–83.

26
Hepatic Cancer

26.1 Definition

Primary tumors of the liver encompass seven categories:

1. *Epithelial tumors*: Benign: hepatocellular adenoma (liver cell adenoma) and focal nodular hyperplasia. Malignant: hepatocellular carcinoma (HCC, liver cell carcinoma), combined hepatocellular and cholangiocarcinoma, hepatoblastoma, and undifferentiated carcinoma.
2. *Mesenchymal tumors*: Benign: angiomyolipoma, lymphangioma and lymphangiomatosis, hemangioma, and infantile haemangioendothelioma. Malignant: epithelioid hemangioendothelioma, angiosarcoma, embryonal sarcoma (undifferentiated sarcoma), and rhabdomyosarcoma.
3. *Miscellaneous tumors*: Solitary fibrous tumor, teratoma, yolk sac tumor (endodermal sinus tumor), carcinosarcoma, Kaposi sarcoma, and rhabdoid tumor.
4. *Hematopoietic and lymphoid tumors*
5. *Secondary tumors*
6. *Epithelial abnormalities*: Liver cell dysplasia—large cell type/small cell type, and dysplastic nodules (adenomatous hyperplasia)—low-grade/high-grade (atypical adenomatous hyperplasia).
7. *Miscellaneous lesions*: Mesenchymal hamartoma, nodular transformation (nodular regenerative hyperplasia), and inflammatory pseudotumor.

Hepatocellular carcinoma (HCC) represents about 75% of all primary liver cancers. Further, hepatoblastoma is a malignant tumor of children, accounting for 79% of all primary liver cancers under the age of 15.

26.2 Biology

Located in the upper right-hand portion of the abdominal cavity, beneath the diaphragm, and on top of the stomach, right kidney, and intestines, the liver is a dark reddish-brown, soft, highly vascular, and easily friable organ that weighs about 1.5 kg and has a prism or wedge shape.

Anatomically, the liver is divided into a larger right lobe and a smaller left lobe by the falciform ligament, each consisting of eight segments (with 1,000 lobules/small lobes). The lobules are connected to small ducts (tubes) that connect with larger ducts to form the common hepatic duct. The bile produced by the liver cells is transferred via the common hepatic duct transports to the gallbladder and via the common bile duct to the duodenum.

The main functions of the liver include production of bile, certain proteins, cholesterol; conversion of excess glucose into glycogen for storage (glycogen can be later reverted to glucose for energy); regulation of blood levels of amino acids; processing of hemoglobin for use of its iron content (the liver stores iron); conversion of poisonous ammonia to urea; clearing the blood of drugs and other poisonous substances; regulating blood clotting; resisting infections by making immune factors and removing bacteria from the bloodstream; and clearance of bilirubin.

Microscopically, the surface of the liver is covered by visceral peritoneum (serosa), with a Glisson capsule underneath. Each hexagonal lobule has a central portal tract with branches of the hepatic artery, the portal vein, and bile ducts, as well as a peripheral tributary of the hepatic vein. Sinusoids are large-diameter capillaries lined by endothelial cells between rows of plates or cords of hepatocytes (which make up 70%–80% of the liver mass), together with Kupffer cells of the reticuloendothelial system.

HCC affects hepatocytes and evolves possibly from hepatocellular adenoma, dysplastic foci, or dysplastic nodules in the cirrhotic liver. Hepatocellular adenoma predominantly occurs in females using oral contraceptives, and also females with maturity onset diabetes of the young type 3, as well as males with glycogen storage disease or androgen treatment. While dysplastic foci are clusters of <1 mm in size without showing invasive growth, dysplastic nodules are clusters of >1 mm in size that show minimal abnormality with a normal to slightly increased nuclear/cytoplasmatic ratio, minimal atypia, no mitoses and one to two cells wide cell plates in low grade lesions; but display increased nuclear/cytoplasmatic ratio, nuclear hyperchromasia, irregular nuclear borders, peripheral location of the nucleus, occasional mitoses, cell plates >2 cells, basophilic cytoplasm, pseudo gland formation and resistance to iron accumulation in high grade lesions [1].

On the other hand, hepatoblastoma appears to derive from pluripotent stem cells of the cirrhosis-free liver.

26.3 Epidemiology

Primary hepatic cancer ranks as the second leading cause of cancer death (after lung cancer) worldwide, with an estimated 782,500 new cases and

745,500 deaths recorded in 2012. Accounting for about 75% of all primary liver cancers, HCC represents the sixth most common type of cancer and is highly prevalent in sub-Saharan Africa, Southeast Asia, and the eastern Mediterranean. HCC shows a predilection for males and is most frequently diagnosed among people of 55–65 years old [1]. Hepatoblastoma usually affects children younger than 3 years.

26.4 Pathogenesis

Risk factors for HCC include hepatitis B virus (54% of cases, mostly in developing countries), hepatitis C virus (31% of cases, mostly in developed countries), diabetes, obesity, aflatoxin exposure, hemochromatosis, severe alcohol intake, and metabolic diseases. These factors induce a perpetual state of inflammation and fibrogenesis, leading to fibrosis, cirrhosis, preneoplastic conditions, and ultimately HCC. In children, Beckwith–Wiedemann syndrome, familial adenomatous polyposis, trisomy 18, and low birth weight are associated with hepatoblastoma; whereas progressive familial intrahepatic cholestasis is linked to HCC [1].

Various genetic alterations (e.g., genomic instability, single-nucleotide polymorphisms, somatic mutations, and deregulated signaling pathways) are implicated in the initiation and progression of HCC as well as hepatoblastoma [2,3].

Notable genomic instability relates to gains of chromosomes 1q21, 1q21-23, 1q21-q22, 1q21.1-q23.2, q22-23.1, 1q24.1-24.2, 8q24.21-24.22, 8q21.13, 8q22.3, and 8q24.3 and losses of Chromosomes 4q34.3-35, 4q13.3-q35.2, 8p, 8p22-p23, D4S2964, and 6q26-q27.

Common single-nucleotide polymorphisms (SNPs) consist of rs2880301 (*TPTE2*), rs17401966 (*KIF1B*), rs17401966 (*KIF1B*), rs455804 (*GRIK1*), rs9272015 (*HLA-DQA1/DRB1*), rs2596542 (*MICA*), rs9275319 (*HLA-DQ*), rs1012068 (*DEPDC5*), rs2551677 (*DDX18*), rs763110 (*FasL*), rs3816747 (*DLC1*), rs7574865, and rs3761549 (*FOXP3*).

Important somatic mutations and deregulated signaling pathways concern *TERT* promoter (telomere stability), *TP53* (cell cycle control), *CTNNB1 and AXIN1* (Wnt/β-catenin signaling), *ARID1A* and *ARID2* (chromatin remodeling), *NFE2L2* and *KEAP1* (oxidative stress), *RPS6KA3* (RAS/MAPK signaling), *JAK1* (JAK/STAT pathway), and *PI3K/AKT* (PI3K-AKT-mTOR pathway) [2,3].

26.5 Clinical features

Symptoms of hepatic cancer may include a lump or pain in the right side below the rib cage, swelling of the abdomen, yellowish skin, easy bruising, weight loss, and weakness. HCC often manifests with abdominal mass, abdominal pain, emesis, anemia, back pain, jaundice, itching, weight loss, and fever. Hepatoblastoma may cause hemihyperplasia (10% of cases), isosexual precocity, midface hypoplasia and slitlike indentations of the earlobe (late features of Beckwith–Wiedemann syndrome), anorexia (in advanced disease), and other clinical signs.

26.6 Diagnosis

Early HCC (also called small HCC) is a well-differentiated lesion of <2 cm in size, with poorly defined margins and vaguely nodular type (which occurs more often in cirrhosis). Progressed HCC is a lesion of >2 cm in size or <2 cm size with moderately differentiated, distinctly nodular, massive, or diffuse type. The nodular type is either a single or multiple nodules; the massive type is a large tumor with irregular demarcation; and the diffuse type contains many small nodules in a liver lobe or the whole organ. HCC growing extrahepatically with a peduncle is called "pedunculated"; if a peduncle is absent, it is called "protruding."

In gray-scale ultrasonography, low-differentiated HCC foci of <3 cm typically appear as hypoechoic lesions, with some showing increased echogenicity due to inclusion of fatty tissue. In contrast-enhanced ultrasonography, benign hepatic lesions are usually hyper- or isoechoic to the surrounding liver parenchyma, while malignant hepatic lesions show hypoechogenicity. On MRI, large HCC displays decreased signal intensity in T1-weighted and increased signal intensity in T2-weighted images, although small HCC may remain isointense to the adjacent liver parenchyma in T1-weighted images [4].

Histologically, HCC is typically a well-vascularized tumor with wide trabeculae (more than three cells), prominent acinar pattern, small cell changes, cytologic atypia, mitotic activity, vascular invasion, absence of Kupffer cells, and the loss of the reticulin network. The common histologic growth patterns of HCC range from trabecular-resembling normal liver tissue, pseudoglandular or acinar with possible bile or fibrin content, and solid or compact pattern. The tumor cells often show Mallory bodies and pale bodies [5].

Early HCC of the vaguely nodular type has a reduced density of unpaired arteries compared to progressed HCC and therefore appears hypovascular in imaging. Early HCC of the distinctly nodular type as well as progressed HCC appear hypervascular because of earlier neovascularization with unpaired arteries. Progressed HCC with classical unpaired arteries are positive for SMA and CD34.

HCC is often graded as I–IV using the four-scale Edmondson and Steiner system. Grade I HCC consists of small tumor cells, arranged in trabeculae, with abundant cytoplasm and minimal nuclear irregularity that are almost indistinguishable from normal liver tissue. Grade II HCC has prominent nucleoli, hyperchromatism, and some degree of nuclear irregularity. Grade III HCC shows more pleomorphism than Grade II and has angulated nuclei. Grade IV HCC has prominent pleomorphism and often anaplastic giant cells.

Histologic variants of HCC include fibrolamellar, sarcomatous, scirrhous, clear cell, steatohepatic HCC, as well as HCC with lymphoid stroma and combined hepatocellular–cholangiocarcinoma. Immunohistochemically, HCC is positive for hepatocyte paraffin-1 (Hep Par-1), GPC-3, CD34, polyclonal carcinoembryonic antigen (pCEA), CD10, arginase-1, heat shock protein-70, glutamine synthetase, CK8, and CK18 but negative for CK7, CK19, and CK20. Tumor markers (e.g., alpha-fetoprotein) may be also employed to confirm HCC in a blood sample. It should be noted that Hep Par-1, CD10, arginase-1, and pCEA do not allow discrimination between benign and malignant hepatocellular tumors. Polyclonal CEA (which produces a distinct "chicken-wire fence" pattern around the canaliculi in HCC) or Hep Par-1 (cytoplasmatic granular positivity in well-differentiated HCC, but <50% in poorly differentiated HCC) may be used to prove the hepatic origin in dealing with suspected liver metastasis [4,5].

Hepatoblastoma is usually a solitary, solid, well-circumscribed, partially encapsulated, variegated, tan-green mass of 10 cm (range 3–20 cm) located in the right lobe. Histologically, the tumor shows epithelial and mesenchymal elements in varying proportions and at variable stages of differentiation. The epithelial type (56%) may present as fetal pattern (31%), embryonal pattern (19%), macrotrabecular pattern (3%), and small cell undifferentiated/anaplastic pattern (3%); while the mixed epithelial mesenchymal type (44%) consists of mixture of fetal/epithelial and mesenchymal cell types. Immunohistologically, the tumor is positive for alpha fetoprotein (negative in small cell type), chromogranin (fetal, epithelial subtypes, usually focal), CK8/18 (fetal, epithelial subtypes), CK19 (embryonal subtypes), EMA, HepPar1 (negative in small cell type), polyclonal CEA (canalicular pattern), and vimentin; but negative for CD45, desmin and neurofilament.

26.7 Treatment

Resection is the treatment of choice for HCC patients with sufficient hepatic function reserve and without advanced fibrosis and portal hypertension. Liver transplantation (LT) is the preferred treatment for patients with early HCC, with evidence of portal hypertension and/or hepatic dysfunction (who are therefore ineligible for resection). Radiotherapy and systemic chemotherapy are not often used in HCC due to intolerance and inefficiency, although local chemotherapy may be used in a procedure. In addition, chemotherapy (e.g., cisplatin, vincristine, cyclophosphamide, and doxorubicin) may be used before and after surgery and transplant [6,7]. Treatment options for hepatoblastoma include preoperative chemotherapy and surgery, or liver transplant if unresectable [8].

26.8 Prognosis

HCC shows recurrence rates of about 50% during the first 3 years and >70% during the first 5 years after surgical resection. The 5-year survival rate after resection for early stage HCC ranges between 17% and 53%. Patients undergoing liver transplantation have an overall survival rate of 75% at 4 years and a recurrence rate of 8%–15%. Hepatoblastoma has a long term survival rate of 60-70% with most recurrences detected within 3 years.

References

1. Cong WM, Bu H, Chen J, et al. Practice guidelines for the pathological diagnosis of primary liver cancer: 2015 update. *World J Gastroenterol.* 2016;22(42):9279–87.
2. Inokawa Y, Inaoka K, Sonohara F, et al. Molecular alterations in the carcinogenesis and progression of hepatocellular carcinoma: Tumor factors and background liver factors. *Oncol Lett.* 2016;12(5):3662–8.
3. Niu ZS, Niu XJ, Wang WH. Genetic alterations in hepatocellular carcinoma: An update. *World J Gastroenterol.* 2016;22(41):9069–95.
4. Nowicki TK, Markiet K, Szurowska E. Diagnostic imaging of hepatocellular carcinoma - A pictorial essay. *Curr Med Imaging Rev.* 2017;13(2):140–53.
5. Schlageter M, Terracciano LM, D'Angelo S, et al. Histopathology of hepatocellular carcinoma. *World J Gastroenterol.* 2014;20(43): 15955–64.
6. Best J, Schotten C, Theysohn JM, et al. Novel implications in the treatment of hepatocellular carcinoma. *Ann Gastroenterol.* 2017;30(1):23–32.
7. PDQ Adult Treatment Editorial Board. *Adult Primary Liver Cancer Treatment (PDQ®): Health Professional Version.* PDQ Cancer Information Summaries [Internet]. Bethesda (MD): National Cancer Institute (US); 2002–2017.
8. PDQ Pediatric Treatment Editorial Board. *Childhood Liver Cancer Treatment (PDQ®): Health Professional Version.* PDQ Cancer Information Summaries. Bethesda, MD: National Cancer Institute (US); 2002–2016.

27
Pancreatic Cancer

27.1 Definition

The pancreas is a glandular organ with two functional compartments: the digestive enzyme-secreting exocrine pancreas and the hormone-secreting endocrine pancreas.

Tumors affecting the exocrine pancreas range from *benign epithelial tumors* (serous cystadenoma, mucinous cystadenoma, intraductal papillary-mucinous adenoma, mature teratoma), *borderline epithelial tumors* (mucinous cystic neoplasm with moderate dysplasia, intraductal papillary-mucinous neoplasm with moderate dysplasia, solid-pseudopapillary neoplasm), *malignant epithelial tumors* (ductal adenocarcinoma, mucinous noncystic carcinoma, signet-ring cell carcinoma, adenosquamous carcinoma, undifferentiated [anaplastic] carcinoma, undifferentiated carcinoma with osteoclast-like giant cells, mixed ductal-endocrine carcinoma, serous cystadenocarcinoma, mucinous cystadenocarcinoma—noninvasive/invasive, intraductal papillary-mucinous carcinoma—noninvasive/invasive [papillary-mucinous carcinoma], acinar cell carcinoma, acinar cell cystadenocarcinoma, mixed acinar–endocrine carcinoma, pancreatoblastoma, solid pseudopapillary carcinoma), *nonepithelial tumors*, to *secondary tumors* [1]. Out of these, pancreatic ductal adenocarcinoma (PDAC, 90%), acinar cell carcinoma (1%–2%), pancreatoblastoma (<1%), and solid pseudopapillary neoplasm (1%–2%) account for about 95% of all pancreatic neoplasms. For this reason, exocrine pancreatic cancer is often referred to as *pancreatic cancer*, which will be the focus of this chapter.

Tumors affecting the endocrine pancreas are neuroendocrine tumors (formerly known as *islet cell tumors*), which comprise glucagonoma, insulinoma, somatostatinoma, gastrinoma, VIPoma, caricinoid, etc., and together account for about 5% of all pancreatic neoplasms. Pancreatic neuroendocrine tumors (PanNET) may be classified as well-differentiated Grade 1 (<2 mitoses per 10 HPF; Ki-67 labeling index <3%), Grade 2 (2–20 mitoses per 10 HPF; Ki-67 labeling index 3%–20%) (containing small to medium-sized ovoid nuclei, minimal pleomorphism, and absence of extensive necrosis), or poorly differentiated Grade 3 (also known as *pancreatic neuroendocrine carcinoma* or PanNEC; >20 mitoses per 10 HPF or Ki-67 index >20%) (containing cytological atypia, apparent pleomorphism, and extensive necrosis) [1]. Since these tumors are detailed in sister volume covering the endocrine system, they will be merely mentioned here as differential diagnoses for pancreatic cancer.

27.2 Biology

Situated on the posterior abdominal wall behind the stomach, the pancreas is a J-shaped (like a hockey stick), flattened, soft, lobulated organ of 12–15 cm in length, consisting of five regions (head, uncinate process, neck, body, and tail).

Functionally, the pancreas is divided into two compartments: exocrine and endocrine. The exocrine pancreas (serous gland) represents >95% of the pancreatic parenchyma and includes a million "berry-like" clusters of zymogenic cells (acini, which secrete digestive enzymes), connected by ductules with associated connective tissue, vessels, and nerves. The ductules join to form intralobular ducts, which then form interlobular ducts that drain into branches of the main pancreatic duct. The main pancreatic duct unites with the common bile duct, forming the hepatopancreatic ampulla of Vater, which opens into the duodenum and controls the flow of the bile and pancreatic fluid with a muscular valve (the sphincter of Oddi). Under the influence of secretin and cholecystokinin, the zymogenic cells secrete trypsin (digesting proteins), lipase (digesting fats), amylase (digesting carbohydrates), and other enzymes, whereas ductular cells produce bicarbonate, rendering the pancreatic fluid alkaline.

The endocrine pancreas represents about 2% of pancreatic parenchyma and includes pancreatic islets (clusters) (or *islets of Langerhans*) scattered throughout the pancreas. The islets contain several cell types (alpha, beta, delta, A, B, C, D, E, F), of which alpha cells (20% of islets) secrete glucagon; beta cells (70% of islets) secrete insulin and amylin; delta cells (<10% of islets) secrete somatostatin; A cells (as well as D cells) secrete adrenocorticotropin, serotonin (5-HT), melanocyte-stimulating hormone (MSH), and vasoactive intestinal peptide; D cells secrete gastrin; E cells (or epsilon cells, <1% of islets) secrete ghrelin; and F cells (or PP/gamma cells, <5% of islets) secrete pancreatic polypeptide. The secreted hormones are transferred via the blood to control energy metabolism and storage throughout the body.

The normal ductal and ductular epithelium is a cubiodal to low-columnar epithelium with amphophilic cytoplasm, but without mucinous cytoplasm, nuclear crowding, and atypia that are commonly seen in neoplastic transformation. PDAC arises from the ductal cells of the exocrine pancreas, particularly in the head region. Precursor lesions for PDAC include pancreatic intra epithelial neoplasia (PanIN), mucinous cystic neoplasm (MCN), and intraductal papillary mucinous neoplasms (IPMN). As a more common precursor lesion, PanIN includes three stages (1 to 3). PanIN stage 1 is a flat epithelial lesion composed of tall columnar cells with basally located nuclei, abundant supranuclear mucin and slight nuclear atypia; PanIN stage 2 may be a flat or papillary lesion showing loss of polarity, nuclear crowding, enlarged nuclei, pseudo-stratification, and hyperchromatism; PanIN stage 3 is usually a papillary or micropapillary lesion showing true cirbriforming, the appearance of 'budding off' of small clusters of epithelial cells into the lumen, and luminal necrosis. MCN is a large mucin producing columnar epithelial cystic lesion supported by ovarian type stroma usually found in the body and tail of the pancreas. IPMN develops in the main pancreatic duct or its major branches, and resembles PanIN at cellular levels but grows into large cystic structures [2]. Acinar cell carcinoma and pancreatoblastoma evolve from acinar cells [1].

27.3 Epidemiology

Pancreatic cancer is the thirteenth most common cancer, with 200,000 new cases per year and an annual incidence rate of 8–12 per 100,000 worldwide. PDAC alone makes up 90% of all pancreatic cancer and is the most lethal form of the disease.

Pancreatic cancer mainly affects people of >45 years, with increasing incidence in older ages (80% patients aged 70–80 years; 1% <40 years). There is a slight male bias (male-to-female ratio, 1.3:1), especially in black males (14.8 per 100,000 vs. 8.8 per 100,000 in the general population). Maoris in New Zealand and native Hawaiians are also vulnerable to pancreatic cancer.

27.4 Pathogenesis

Risk factors for pancreatic cancer include cigarette smoking; obesity; alcohol; coffee; high fat, high protein, low fiber diet; chronic pancreatitis; diabetes mellitus; and syndromic familial pancreatic cancer (e.g., hereditary pancreatitis [PRSS1], hereditary nonpolyposis colorectal cancer [HNPCC or Lynch 11; hMSH2, hMLH1], familial breast cancer [BRAC-2], familial atypical multiple mole melanoma [FAMMM syndrome; P16], and familial adenomatous polyposis [FAP; APC]).

Molecularly, pancreatic cancer is linked to gene mutations, aberrant DNA methylation, and altered mitochondrial DNA. The most common genetic alterations in pancreatic cancer involve oncogenes (*KRAS* [chromosome 12p, point mutations in codon 12, 90% of cases], *BRAF* [chromosome 7q, point mutations, tumors with wild-type *KRAS*], *MYB* [chromosome 6q, amplification, 10%], *AKT2* [chromosome 19q, amplification, 10-20%], *AIB1* [chromosome 20q, amplification, 66%], *HER/2-neu* [chromosome 17q, overexpression, large range]), tumor suppressor genes (*P16/CDKN2A* [chromosome 9p, loss of heterozygosity/intragenic mutation/homozygous mutation/hypermethylation, 95% of cases], *TP53* [chromosome 17p, loss of heterozygosity/intragenic mutation, 50–75%], *MAD4/DPC4* [chromosome 18q, loss of heterozygosity/intragenic mutation/homozygous mutation, 55%], EP300 [chromosome 22q, loss of heterozygosity/ intragenic mutation, 25%], *MKK4* [chromosome 17p, loss of heterozygosity/ intragenic mutation/homozygous mutation, 4%], *TGFβR1 (ALK 5)* [chromosome 2q, homozygous mutation, 2%], *TGFβR2* [chromosome 3p, homozygous mutation/bi-allelic intragenic mutations, 4–7%],), DNA mismatch repair genes (*MLH1* [chromosome 3p, germline/hypermethylation, 3–15%], *BRCA2* [chromosome 13q, germline, 7%]), and mitochondrial genome [mitochondrial DNA, intragenic mutations, 100%]) [2,3].

27.5 Clinical features

Early stage pancreatic cancer does not usually produce noticeable symptoms, whereas late stage pancreatic cancer may induce jaundice, light-colored stools, dark urine, pain in the upper or middle abdomen and back, weight loss, loss of appetite, and fatigue.

27.6 Diagnosis

PDAC (also known as *infiltrating ductal adenocarcinoma*) is a poorly demarcated, firm white-yellow mass with the adjacent non-neoplastic pancreas appearing atrophic and fibrotic and the pancreatic ducts being dilated. On CT, PDAC typically presents as a poorly defined mass with extensive surrounding desmoplastic reaction, and appears hypodense on arterial phase scans in 75–90% of cases, but may become isodense on delayed scans. A PDAC that surrounds a vessel by >180 degrees as shown by CT is deemed T4 disease and unresectable. On MRI, PDAC appears hypointense on T1 weighted images, and variable (depending on the amount of desmoplastic reaction) on T2 weighted images, and displays slower enhancement on T1 + C (Gd). Microscopically, PDAC shows glandular and ductal structures, abundant desmoplastic stroma, eosinophilic to clear cytoplasm, enlarged pleomorphic nuclei, perineural, lymphatic and blood vessel invasion. Depending on the level/extent of glandular differentiation, PDAC can be distinguished into three grades. Grade 1 has well-differentiated duct-like glands, intense mucin production, little nuclear polymorphism and polar arrangement, and <5 mitoses per 10 HPF; grade 2 has moderately differentiated duct-like structures and tubular glands, irregular mucin production, moderate nuclear polymorphism, and 6-10 mitoses per 10 HPF; and grade 3 has poorly differentiated glands, muco-epidermoid and pleomorphic structures, abortive mucin production, marked nuclear pleomorphism and increased nuclear size, and >10 mitoses per 10 HPF [2]. Immunohistochemically, PDAC is positive for CEA (90%), mesothelin (95%), B72.3 (92%), mucin (MUC1 [pan-epithelial mucin], MUC3, MUC4, and MUC5AC [gastric foveolar mucin]) and carbohydrate antigens (e.g., CA19-9) and displays aberrant TP53 expression or SMAD4 loss [1,2].

Notable features for establishing a diagnosis of pancreatic cancer include: (i) pancreatic carcinoma infiltrates in a haphazard pattern; (ii) neoplastic glands occur adjacent to muscular arteries without intervening pancreatic parenchyma, in contrast to the non-neoplastic pancreas, in which muscular arteries run at the periphery of the lobules separated from the ducts by pancreatic parenchyma; (iii) perineural invasion, which is virtually diagnostic of invasive pancreatic cancer; (iv) vascular invasion along the intimal surface of the vessel, with malignant glands lining the intima of the vessel mimicking a low-grade pancreatic intraepithelial neoplasia (PanIN) lesion; (v) the '4 to 1 rule', with the finding of nuclei in a single gland varying in area by more than 4 to 1 indicative of a malignancy; (vi) necrotic debris within the lumen of a gland; (vii) incomplete lumina, i.e., lumina are not completely lined by an epithelial layer and the luminal contents thus appear to directly touch the stroma; (viii) a gland directly touching fat without intervening stroma; (ix) the immunolabeling profile, with the tumor being negative (55%) for the DPC4/MADH4 gene product (dpc4) and positive carcinoembryonic antigen (CEA) and mesothelin, in contrast to non-neoplastic glands's intact dpc expression and frequent negativity for CEA or mesothelin [2].

Adenosquamous carcinoma is a malignant epithelial neoplasm of extreme aggression with significant components of both glandular and squamous differentiation.

Signet ring cell carcinoma is a malignant epithelial neoplasm showing predominantly infiltrating round non-cohesive (isolated) cells containing intracytoplasmic mucin. Before confirmation of a primary pancreatic signet ring carcinoma, there is a need to rule out metastases from a breast or gastric source.

Undifferentiated carcinoma is an epithelial neoplasm of extreme aggression and uniform lethality with pleomorphic epithelioid mononuclear cells containing abundant eosinophilic cytoplasm admixed with bizarre frequently multinucleated giant cells, or relatively monomorphic spindle cells. The absence of obvious glandular structures or other features makes identification of a definite direction of differentiation impossible.

Undifferentiated carcinoma with osteoclast-like giant cells is a malignant epithelial neoplasm with reactive multinucleated giant cells admixed with atypical neoplastic mononuclear cells, and often arises in association with an adenocarcinoma or mucinous cystic neoplasm.

Acinar cell carcinoma is a large solitary, solid and well-circumscribed mass showing acinar cell differentiation (similar to pancreatoblastoma). The cyanophilic acinar appearing cells arranged in sheets and trabecular pattern contain granular cytoplasm and centrally located nucleus with prominent nucleolus. It is rare neoplasm with poor prognosis.

Pancreatoblastoma is a solid and cystic lesion that contains neoplastic cells arranged in sheets, tubules, nests and acinar patterns, and shows acinar differentiation, squamoid nests (which help differentiate from acinar cell carcinoma), neuroendocrine and primitive blastemal histomorphology.

Solid-pseudopapillary neoplasm is a solid and cystic mass with extensive degenerative changes and uncertain differentiation. It shows poorly cohesive uniform cells containing eosinophilic or clear vacuolated cytoplasm, round to oval nuclei, often grooved or indented, along with eosinophilic globules and foamy macrophages.

Differential diagnoses for PDAC cancer include pancreatic neuroendocrine tumors (mean 58 years vs PDAC's 66 years, solid or cystic lesion with endocrine differentiation, nested or trabecular growth pattern, distinct neuroendocrine morphology [granular amphophilic to eosinophilic cytoplasm and coarsely clumped "salt and pepper" chromatin], and positivity for neuroendocrine markers [synaptophysin and chromogranin A] and peptide hormones [e.g., insulin and glucagon]), pancreatitis, pancreatic atrophy, pancreatic lipomatosis, and pseudopancreatitis.

27.7 Treatment

Surgical resection is curative for small, localized exocrine pancreatic cancer but generally not for unresectable, metastatic, or recurrent disease. Being resistant to treatment with chemotherapy and radiation, pancreatic tumor symptoms may be relieved with palliative measures, which improve quality of life but not overall survival [4–6].

27.8 Prognosis

Exocrine pancreatic cancer (PDAC) is rarely curable, with a median survival time of 6 months after diagnosis, and a mortality rate of 95%–98% after 1 year [7].

References

1. Rishi A, Goggins M, Wood LD, Hruban RH. Pathological and molecular evaluation of pancreatic neoplasms. *Semin Oncol.* 2015;42(1):28–39.
2. Hruban RH, Fukushima N. Pancreatic adenocarcinoma: update on the surgical pathology of carcinomas of ductal origin and PanINs. *Mod Pathol.* 2007;20 Suppl 1:S61–70.
3. Hackeng WM, Hruban RH, Offerhaus GJ, Brosens LA. Surgical and molecular pathology of pancreatic neoplasms. *Diagn Pathol.* 2016;11(1):47.
4. PDQ Adult Treatment Editorial Board. *Pancreatic Cancer Treatment (PDQ®): Health Professional Version.* PDQ Cancer Information Summaries. Bethesda, MD: National Cancer Institute (US); 2002–2016.
5. Fotopoulos G, Syrigos K, Saif MW. Genetic factors affecting patient responses to pancreatic cancer treatment. *Ann Gastroenterol.* 2016;29(4):466–76.
6. Polireddy K, Chen Q. Cancer of the pancreas: Molecular pathways and current advancement in treatment. *J Cancer.* 2016;7(11):1497–514.
7. Weledji EP, Enoworock G, Mokake M, Sinju M. How grim is pancreatic cancer? *Oncol Rev.* 2016;10(1):294.

28
Small Intestine Cancer

Andreas V. Hadjinicolaou and Christopher Hadjittofi

28.1 Definition

Neoplasms affecting the small intestine include primary benign tumors (adenoma, leiomyoma, fibroma, and lipoma), primary malignant tumors (adenocarcinoma, neuroendocrine tumor [NET], lymphoma, and sarcoma), and secondary tumors (lymphoma, melanoma, breast, lung, colon, and kidney tumors). Out of these, the most common histological types of primary small intestine cancer are adenocarcinoma (accounting for about 30% of small intestine malignancies), NET (or carcinoid tumor, 40%), lymphoma (20%), and sarcoma (10%).

28.2 Biology

The small intestine (small bowel) is part of the gastrointestinal tract that also includes the esophagus, large intestine, and stomach. Connecting between the stomach and the large intestine, the small intestine is a convoluted tube about 6 m long, which makes up 75% of the length and 90% of the absorptive surface area of the gastrointestinal tract. The small intestine has three sections: the duodenum, jejunum, and ileum.

The wall of the small intestine consists of four layers: mucosa (epithelium, lamina propria, muscularis mucosa), submucosa (connective tissue, blood vessels, lymphatics, submucosal or Meissner's plexus), muscularis propria (externa) (inner circular and outer longitudinal layer), and serosa (mesothelial lining, loose connective tissue).

The main functions of the small intestine are to further degrade food with enzymes secreted in the lumen or transported from the pancreas and liver and to absorb nutrients through the intestinal wall into the bloodstream.

Adenocarcinoma develops in the gland cells of the small intestine. NET (or carcinoid tumor) arises in the hormone-producing cells of the ileum and, rarely, the duodenum. Lymphoma is usually found in the ileum and jejunum and is most commonly non-Hodgkin lymphoma. Sarcoma is a type of mesenchymal tumor that occurs throughout the small intestine, with leio-myosarcoma and gastrointestinal stromal tumors being the most common.

28.3 Epidemiology

Primary malignancies of the small intestine account for 2%–3% of digestive system tumors and 0.42% of total cancer cases, with global incidence of <1.0 per 100,000. In the United States, the annual incidence of adenocarcinoma in men and women during 1992–2006 was 1.45 and 1.00 per 100,000, respectively; that of carcinoid tumor was 1.00 and 0.70 per 100,000, respectively; that of lymphoma was 0.54 and 0.26 per 100,000, respectively; and that of sarcoma was 0.24 and 0.17 per 100,000, respectively [1].

The average age at diagnosis is 65–66 years for all small intestine cancers. Whereas adenocarcinoma and carcinoid present at 67–68 years, lymphoma and sarcoma present at 60–62 years. Men have higher rates of all types of small intestine cancer than women, with a male-to-female ratio of 1.4:1.

28.4 Pathogenesis

Adenocarcinoma usually develops from adenomas due to genetic altera-tions that resemble the adenoma-carcinoma sequence of colorectal cancer [2]. These include mutations and/or overexpression of *TP53, APC, CTNNB1* (β-catenin), *HER2 (ERBB2), KRAS, VEGF-A, EGFR,* and deficiency in mismatch repair (dMMR) genes. Familial adenomatous polyposis, Lynch syn-drome (hereditary nonpolyposis colorectal cancer), hamartomatosis (related to Peutz–Jeghers, juvenile polyposis, or Cowden's syndrome), Crohn's dis-ease, celiac disease, and cystic fibrosis are also linked to the pathogenesis of adenocarcinoma. Other risk factors include alcohol intake, smoking, and consumption of red meat, sugar, and starch [3].

NET has been associated with multiple endocrine neoplasia (MEN1 and MEN4) syndromes. Alterations implicated in NET range from single nucle-otide variants (at the *VHL, BRAF, FGFR2, MEN1, MLF1, SRC, SMAD,* and *FANCD2* genes); chromosomal gain (4, 5, 19, 20) or loss (11, 18); cancer-related pathways (PI3K/Akt/mTOR and TGF-β); cell cycle gene mutations (e.g., *CDKN1B*); neurodevelopmental transcription factors; and dysregula-tion of protein expression via microRNA alterations [4].

28.5 Clinical features

The principal symptom of small intestine cancer is chronic intermittent cramp-like abdominal pain, often with nausea and vomiting, melena, diarrhea, fatigue, anorexia, and unexplained weight loss. Iron-deficiency anemia through occult GI bleeding is often the only detectable abnormality. On rare occasions, patients may present acutely with intestinal obstruction, with or without an appreciable abdominal mass or even perforation.

The frequencies of clinical features in small intestine adenocarcinoma are abdominal pain (42%–83%), weight loss (23%–87%), abdominal mass (19%–29%), anemia (18%–75%), nausea/vomiting (27%–34%), bleeding (13%–68%), obstruction (16%–65%), jaundice (18%–30%), and anorexia (18%–25%) [5]. In the case of carcinoids, serotonin, kinins, and prostaglandins are released, giving rise to the carcinoid syndrome (skin flushing, telangiectasia, diarrhea, cardiac murmurs, tachycardia, dyspnea, and wheeze). Carcinoid syndrome correlates strongly (~90%) with metastatic disease.

28.6 Diagnosis

If small intestine cancer is suspected, a thorough history should be obtained and screening for fecal occult blood undertaken. Basic investigations should be carried out, including complete blood count, serum electrolytes, and liver function tests.

As esophagogastroduodenoscopy is limited to the proximal duodenum, and colonoscopy allows assessment up to the terminal ileum, leaving the small intestine largely out of reach, wireless video capsule endoscopy (VCE) is becoming the investigation of choice for suspected small intestine pathology. Double-balloon enteroscopy, although invasive, complements VCE as it achieves tissue biopsy for histological examination as well as dilatation of strictures, stent placements, or other therapeutic procedures.

Barium small intestine follow-through may reveal masses or intussusception and can visualize the lumen and mucosa. Enteroclysis is similar to but more sensitive than traditional follow-through. However, as these procedures cannot reveal extraintestinal lesions, abdominal contrast-enhanced CT has become the main diagnostic and staging imaging method for SI cancer (especially adenocarcinoma and carcinoids).

Staging of small intestine adenocarcinoma depends on tumor size, lymph node involvement, and distant metastasis according to the TNM system (Tables 28.1 and 28.2) [6].

Carcinoembryonic antigen is often elevated in serum and detected by immunohistochemistry in advanced adenocarcinomas. Carcinoids secrete active hormonal metabolites (e.g., 24-hour urinary excretion of 5-HIAA,

Table 28.1 TNM Staging of Small Intestine Adenocarcinoma

Primary Tumor Assessment (T)		Regional Lymph Nodes Assessment (N)		Distant Metastasis Assessment (M)	
Tx	Primary tumor size cannot be evaluated	Nx	Regional lymph nodes cannot be evaluated	M0	No distant metastasis
T0	No evidence of primary tumor	N0	No regional lymph node involvement	M1	Distant metastasis present
Tis	Carcinoma in situ (the cancer is only seen in the epithelium without penetration of deeper tissue	N1	Metastatic involvement of 1-3 regional lymph nodes		
T1	Tumor invades the lamina propria or muscularis mucosa (T1a) or submucosa (T1b)	N2	Metastatic involvement of 4 or more regional lymph nodes		
T2	Tumor invades the muscularis propria				
T3	Tumor invades into the subserosa or into the mesentery or retroperitoneum by kess than 2 cm				
T4	Tumor invades through the visceral peritoneum to directly affect other organs or structures such as: • other loops of the small intestine, mesentery or retroperitoneum by more than 2 cm or, • through the serosa into the abdominal wall or pancreas (the latter applies specifically to duodenal tumors) • liver, lungs and other organs				

Table 28.2   Prognosis of Small Intestine Adenocarcinoma by Stage and Grade

Stage	Primary Tumor Size (T)	Regional Lymph Nodes (N)	Distant Metastasis (M)	Prognosis (5-yr Survival)	Proportion of Presenting Patients
0	Tis	N0	M0	N/A	–
1	T1 or T2	N0	M0	55%	0%
2A	T3	N0	M0	49%	43% (stage 2A & 2B)
2B	T4	N0	M0	35%	See above
3A	Any T	N1	M0	31%	17% (stage 3A & 3B)
3B	Any T	N2	M0	18%	See above
4	Any T	Any N	M1	5%	36%

Tumor Grading (G)		Frequency of Tumors	Prognosis (10-year Survival)
Gx	Tumor grade cannot be evaluated	–	–
G1	Well differentiated cancer cells	0–42%	43%
G2	Moderately differentiated cancer cells	24–45%	43%
G3	Poorly differentiated cancer cells	34–42%	–
G4	Undifferentiated cancer cells	34–42%	0%

Sources: The 7th edition of the AJCC Staging Manual 2010; American Cancer Society website; Kummar S, Ciesielski TE, Fogarasi MC. *Management of Small Bowel Adenocarcinoma Oncology (Cancer Network).* 2002;10:1364–9.

serum serotonin, and chromogranin A), which when combined offer highly sensitive and specific diagnosis. Additionally, somatostatin receptor scintigraphy with octreotide can localize carcinoids, which like most NETs often express high levels of somatostatin receptors [4].

28.7 Treatment

Treatment for small intestine cancer depends on the type, site, and stage of tumor and includes surgery, chemotherapy, biological therapy, and radiotherapy [6].

For localized adenocarcinoma, surgical options include pancreatoduodenectomy (proximal duodenum), segmental resection (distal duodenum), wide excision (jejunum or proximal ileum), distal ileal resection, and right hemicolectomy (distal ileum). Additionally, adjuvant chemotherapy is offered after surgery in cases of lymph node involvement. In certain cases, neoadjuvant or adjuvant chemoradiotherapy may also be offered.

For metastatic adenocarcinoma or advanced local adenocarcinoma not amenable to surgery, chemotherapy is only offered to patients with good performance status—otherwise palliative care is recommended. Hepatic metastasectomy is also possible in selected cases.

For localized carcinoid, wide tumor resection with mesenterectomy and lymphadenectomy provides the best prognosis.

For metastatic carcinoid, resection of primary tumor is often reserved for debulking, for the prevention of mesenteric fibrosis, or for symptom control. Hepatic metastasectomy prolongs disease-free survival in patients with advanced disease and is thus often performed in asymptomatic patients with amenable lesions. Octreotide must be administered preoperatively to avoid carcinoid crisis. For patients with carcinoid syndrome and widespread metastatic disease that is nonresectable, hepatic artery embolization provides a nonsurgical alternative in liver metastases. Chemotherapy is ineffective in reducing carcinoid tumor burden. The main systemic treatment for carcinoid syndrome involves somatostatin analogues to relieve the symptoms of hormonal metabolites.

28.8 Prognosis

Prognosis for patients with small intestine adenocarcinomas is generally poor. Patients with well, moderately, and poorly differentiated tumors survive for a median of 66, 40, and 14 months after diagnosis, respectively. The 5-year overall survival rates for patients with Stages I, IIA, IIB, IIIA, IIIB, and IV adenocarcinomas are 55%, 49%, 35%, 31%, 18%, and 5%, respectively (Table 28.2).

Patients with well-differentiated carcinoid tumors have a 5-year survival rate ranging from 36% for those with metastatic jejunal or ileal disease to over 95% for those with localized duodenal disease.

References

1. Siegel RL, Miller KD, Jemal A. Cancer statistics, 2015. *CA Cancer J Clin*. 2015;65(1):5–29.

2. Hadjinicolaou AV, Hadjittofi C, Athanasopoulos PG, Shah R, Ala AA. Primary small bowel melanomas: Fact or myth? *Ann Transl Med.* 2016;4(6):113.

3. Pourmand K, Itzkowitz SH. Small bowel neoplasms and polyps. *Curr Gastroenterol Rep.* 2016;18(5):23.

4. Xavier S, Rosa B, Cotter J. Small bowel neuroendocrine tumors: From pathophysiology to clinical approach. *World J Gastrointest Pathophysiol.* 2016;7(1):117–24.

5. Aparicio T, Zaanan A, Mary F3 Afchain P, Manfredi S, Evans TR. Small bowel adenocarcinoma. *Gastroenterol Clin North Am.* 2016;45(3):447–57.

6. PDQ Adult Treatment Editorial Board. *Small Intestine Cancer Treatment (PDQ®): Health Professional Version.* PDQ Cancer Information Summaries. Bethesda, MD: National Cancer Institute (US); 2002–2017.

Glossary

anaplasia: A term used to describe cancer cells with a total lack of differentiation and with resemblance to original cells either in functions or structures or both; also known as *dedifferentiation* (backward differentiation).

apoptosis: Programmed cell death, with cells that are damaged beyond repair typically dying as they swell and burst, spilling their contents over their neighbors.

atypia: The state of being not typical or normal. In medicine, atypia is an abnormality in cells in tissue, which may or may not be a precancerous indication associated with later malignancy.

benign tumor: A slow-growing, noncancerous tumor that does not invade nearby tissue or spread to other parts of the body. In most cases, a benign tumor has a favorable outcome, with or without surgical removal. However, a benign tumor in vital structures such as nerves and blood vessels, or that is undergoing malignant transformation, often has serious consequences (see hamartoma).

biopsy: A procedure to remove tumor tissue or cells or tissues for microscopic examination. This is usually conducted through excisional biopsy (removal of an entire lump of tissue), incisional biopsy (removal of part of a lump or a sample of tissue), core biopsy (removal of tissue using a wide needle), or fine-needle aspiration biopsy (removal of tissue or fluid using a thin needle).

calcification: The accumulation of calcium salts (e.g., calcium phosphate) in body tissues such as tumor mass, where they do not usually appear. This leads to tissue hardening and produces a dense opacity on a radiographic image.

cancer (plural *cancers* or *cancer*): A group of diseases involving abnormal or uncontrolled expansion of cells that has the potential to invade nearby tissue and/or spread to other parts of the body (see *tumor*, *neoplasm*, *lesion*).

carcinoma: A type of cancer that begins in a tissue (called *epithelium*) that lines the inner or outer surfaces of the body.

CT (computerized tomography): Also known as *computed tomography scan*, *CT scan*, or *computerized axial tomography* (CAT), CT utilizes an X-ray machine linked to a computer together with a dye injected into a vein or swallowed to take a series of detailed pictures of affected organs or tissues in the body from different angles, in order to reveal the precise location and dimension of a tumor.

cyst: A closed capsule or sac-like structure usually filled with liquid, semi-solid, or gaseous material (but not pus, which is considered an abscess). An abnormal formation, cyst on the skin, mucous membranes, and inside palpable organs can be felt as a lump or bump, which may be painless or painful. Whereas cysts due to infectious causes are preventable, those due to genetic and other causes are not. Most cysts are benign (noncancerous).

dedifferentiation: See *anaplasia*.

desmoplasia: The growth of fibrous or connective tissue around a neoplasm, causing dense fibrosis; it is considered a hallmark of invasion and malignancy.

differentiated: A term used to describe how much or how little tumor tissue looks like the normal tissue it came from. Well-differentiated cancer cells look more like normal cells and tend to grow and spread more slowly than poorly differentiated or undifferentiated cancer cells.

dysplasia: The overgrowth of immature cells at the location where the number of mature cells is decreasing. This term is particularly used for when cellular abnormality is restricted to the new tissues.

endoscopy: A thin, tubelike instrument with a light and a lens for checking for abnormal areas inside the body.

FISH (fluorescence *in situ* hybridization): FISH is used for determining the positions of particular genes, for identifying chromosomal abnormalities, and for mapping genes of interest.

grade, grading: A measure of cell anaplasia (reversion of differentiation) in tumor, it is based on the resemblance of the tumor to the tissue of origin. Depending on the amount of abnormality, a tumor is graded as 1 (well differentiated; low grade), 2 (moderately differentiated; intermediate grade), 3 (poorly differentiated; high grade), or 4 (undifferentiated; high grade) (see *stage*).

H&E stain: Combined use of hematoxylin (positively charged) and eosin (negatively charged) to stain nucleic acids (negatively charged) in blue and amino groups in proteins (negatively charged) in pink, respectively.

hamartoma: a benign, tumor-like, focal malformation resulting possibly from a developmental error. Composed of abnormal or disorganized mixture of cells and tissues, hamartoma grows at the same rate at the surrounding tissues, and rarely invades or compresses nearby structures significantly. In contrast, a true benign tumor may grow faster than surrounding tissues and compresses nearby structures. Despite its benign histology, hamartoma may be implicated in some rare, but life-threatening clinical issues such as those associated with neurofibromatosis type 1 and tuberous sclerosis. A nonneoplastic mass (eg, hemangioma) can also arise in this way, contributing to misdiagnosis (see benign tumor).

hyperplasia: A disease associated with an increase in the number of normal-looking cells, leading to an enlarged organ; it is also known as *hypergenesis*.

IHC (immunohistochemistry): A technique that exploits the principle of antibodies binding specifically to antigens in biological tissues to visualize the distribution and localization of specific cellular components within cells and in the proper tissue context.

Ki-67: A nuclear protein (also known as *KI-67* or *MKI67*) associated with cellular proliferation. Ki-67 protein is present during all active phases of the cell cycle (G_1, S, G_2, and mitosis) but is absent in resting cells (G_0). The fraction of and ribosomal RNA transcription. Ki-67 protein is present during all active phases of the cell cycle [G1 (pre-DNA synthesis), S (DNA synthesis), G2 (post-synthesis), and M (mitosis)], but absent in resting phase (G_0). The fraction of Ki-67-positive tumor cells detected by MIB-1 (the Ki-67 labeling index or MIB-1 labeling index) often correlates to the aggressiveness and thus the clinical course of cancer (see MIB-1).

lesion: A term in medicine to describe all abnormal biological tissue changes, such as a cut, a burn, a wound, or a tumor. In cancer, *lesion* is used interchangeably with *tumor*, *cancer*, or *neoplasm* (see *cancer, tumor, neoplasm*).

LOH (loss of heterozygosity): A gross chromosomal event that results in loss of the entire gene and the surrounding chromosomal region.

malignancy: The state or presence of a malignant tumor.

malignant tumor: A tumor with the capability of invading surrounding tissues, producing metastases, and recurring after attempted removal.

metaplasia: The reversible replacement of one differentiated cell type with another mature differentiated cell type.

MIB-1: a monoclonal antibody raised against Ki-67 protein that allows accurate immunohistochemical detection of active or growing cells (see Ki-67).

mitotic figure: Microscopic detection of the chromosomes as tangled, dark-staining threads in cells undergoing mitosis; it is often expressed as mitotic figures per 10 high power fields (hpf, usually 400-fold magnification) (mitotic activity index) or per 1000 tumor cells (mitotic index). As mitotic cell count per 10 hpf equals to an area 0.183 mm^2, the American Joint Committee on Cancer specifies that mitotic rate (the proportion of cells in a tissue undergoing mitosis) be reported as mitoses per mm^2, with a conversion factor of 1 mm^2 equaling 4 hpf.

MRI (magnetic resonance imaging): Also called *nuclear magnetic resonance imaging* (NMRI), MRI relies on a magnet, radio waves, and a computer to take a series of detailed pictures of affected areas inside the body that help pinpoint the location and dimension of tumor mass, if present. MRI has better image resolution than CT. It includes T1-weighted, T2 weighted, fluid attenuated inversion recovery (FLAIR, also called dark fluid technique), and diffusion weighted imaging (DWI) sequences. T1 weighted images reveal anatomical details and information about venous sinus permeability or pathologic blush (e.g., water and CSF appear dark; fat and calcification appear white/gray). Use of intravenous contrast gadolinium in T1-weighted sequences further enhances and improves the quality of the images. T2 weighted images provide information about edema, arteries and sinus permeability (e.g., water appears white/hyperintense; fat and calcification appear gray/dark). FLAIR sequences remove the effects of fluid (which normally covers a lesion) from the resulting images (e.g., CSF appears dark; edema appears enhanced). DWI sequences help visualize acute infarction and other inflammatory lesions.

mutation: A change in the structure of a gene caused by the alteration of single base units in DNA or the deletion, insertion, or rearrangement of larger sections of genes or chromosomes, leading to the formation of a variant that may be transmitted to subsequent generations.

necrosis: A form of cell injury leading to the premature/unprogrammed death of cells and living tissue caused by autolysis (due to too little blood flowing to the tissue; see *apoptosis*).

neoplasia: A term that describes abnormal growth or proliferation of cells, resulting in a tumor that can be cancerous.

neoplasm: A new and abnormal growth of tissue in a part of the body; it is used interchangeably with *tumor* or *cancer*.

oncogene: A gene whose mutation or abnormally high expression can transform a normal cell into a tumor cell.

parenchyma: The functional tissue of an organ as distinguished from the connective and supporting tissue (see *stroma*).

pleomorphism: A term used in histology and cytopathology to describe variability in the size, shape, and staining of cells and/or their nuclei; it is a feature characteristic of malignant neoplasms and dysplasia.

PCR (polymerase chain reaction): a procedure for rapid, in vitro production of multiple copies of particular DNA sequences relevant to diagnosis.

PET (positron emission tomography): A PET scan combines a computer-based scan with a radioactive glucose (sugar) injected into a vein to generate a rotating picture of the affected area, with malignant tumor cells showing up more brightly due to their more active uptake of glucose than normal cells.

radiography: A term used to collectively describe electromagnetic radiation (especially X-rays) based procedures to visualize the internal structure of a non-uniformly composed and opaque object such as the human body (see *MRI, CT*).

radiotherapy: Also called *radiation, radiation therapy*, or *X-ray therapy*, radiotherapy involves delivery of radiation externally through the skin or internally (brachytherapy) for destruction of cancer cells or inhibition of their growth.

stage, staging: As a measurement of the extent to which a tumor has spread, stage (usually ranging from 0, I, II, III to IV) is commonly expressed in two ways: clinical and pathological stages. A clinical stage is an estimate of the extent of the tumor after physical exam, imaging studies (e.g., x-rays, CT, MRI) and tumor biopsies as well as other tests (e.g., blood tests), and provides a vital means for selecting most appropriate treatment and for comparing the tumor response to treatment. A pathological stage relies on the results of the exam and tests mentioned above, in addition to information obtained during surgery (e.g., the actual degree of tumor spread, nearby lymph node involvement), and thus offers a more precise guide than clinical stage in predicting treatment outcome and prognosis (see *grade, TNM*).

stroma: The parts of a tissue or organ that have a connective and structural role and that do not conduct the specific functions of the organ (e.g., connective tissue, blood vessels, nerves, ducts) (see *parenchyma*).

TNM: A system (also known as the *TNM system*) designed by American Joint Committee on Cancer to pathologically stage a solid tumor. T (tumor; TX, T0, T1, T2, T3, T4) indicates the depth of the tumor invasion – the higher the number, the further the cancer has invaded; N (nodes; NX, N0, N1, N2, N3) indicates whether the lymph nodes are affected, and how much the tumor has spread to lymph nodes near original site; M (metastasis; MX, M0, M1) indicates whether the tumor has spread to other parts of the body. For example, the pathological stage of rectal tumor may be designated as stage 0 (carcinoma in situ), stage I (T1 or T2/N0/M0), stage II (T3a, T4a or T4b/N0/M0), stage III (T1-T4/N1-N2/M0), and stage IV (any T/any N/M1), with number after each letter providing further details about the tumor)(see grade, stage).

translocation: a segment from one chromosome is transferred to a non-homologous chromosome (interchromosomal transloaction) or to a new site on the same chromosome (intrachromosomal translocation). A reciprocal translocation occurs when two nonhomologous chromosomes swap parts; whereas a non-reciprocal translocation occurs when the transfer of chromosomal material is one way, i.e., another segment does not exchange places with the first segment.

tumor: A swelling of a part of the body, generally without inflammation, caused by an abnormal growth of tissue, whether benign or malignant (see *cancer, neoplasm, lesion*).

tumor suppressor gene: A gene (also known as *antioncogene*) that regulates cell division, repairs DNA mistakes, or instructs cells when to die (see *apoptosis*). When a tumor suppressor gene is mutated, cell growth gets out of control.

ultrasound: A device for delivering sound waves that bounce off tissues inside the body like an echo and for recording the echoes to create a picture (sonogram) of areas inside the body.

undifferentiated: The presence of very immature and primitive cells that do not look like cells in the tissue of their origin. Undifferentiated cells are said to be anaplastic and an undifferentiated cancer is more malignant than a cancer of that type that is well differentiated (see *anaplasia, differentiated*).

Index

POCKET GUIDES TO
BIOMEDICAL SCIENCES

Series Editor
Dongyou Liu